Differentially Expressed
Genes in Plants

Differentially Expressed Genes in Plants:
A Bench Manual

EDITED BY E. HANSEN AND G. HARPER

UK   Taylor & Francis Ltd, 4 John Street, London WC1N 2ET
USA Taylor & Francis Inc., 1900 Frost Road, Suite 101, Bristol, PA 19007

British Library Cataloguing in Publication Data

A catalogue record for this book is available from the British Library.

ISBN 0-7484-0421-X (cased)

Library of Congress Cataloguing Publication data are available

Cover design by Jim Wilkie

Typeset in Times 10/12pt by Keyword Typesetting Services, Surrey

Printed in Great Britain by T. J. International Ltd, Cornwall

CONTENTS

Contents

Contributors

Guy Bauw and Marc Van Montagu
Laboratorium voor Genetica, Department of Genetics, affiliated to the Flanders Interuniversity Institute for Biotechnology, Universiteit Gent, K.L. Ledeganckstraat 35, B-9000 Gent, Belgium

Ekkehard Hansen
Laboratório de Biotecnologia, Centro de Biociências e Biotecnologia, Universidade Estadual do Norte Fluminense 28015-620, Campos dos Goytacazes-RJ, Brazil

Glyn Harper
Department of Virology, John Innes Centre, Colney Lane, Norwich, NR4 7UH, UK

Gregory R. Heck and Donna E. Fernandez
Department of Botany, University of Wisconsin, 430 Lincoln Drive, Madison, WI 53706-1381, USA

Susanne E. Kohalmi, Jacek Nowak, and William L. Crosby
Gene Expression Group, Plant Biotechnology Institute, National Research Council of Canada, 110 Gymnasium Place, Saskatoon, SK S7N 0W9, Canada

Charles P. Scutt
Centre for Plant Biochemistry and Biotechnology, University of Leeds, Leeds, LS2 9JT, UK

Jennifer F. Topping and Keith Lindsey
Department of Biological Sciences, University of Durham, South Road, Durham, DH1 3LE, UK

Preface

The phenomenon of change is of prime interest to biologists. Alterations in patterns of gene expression within cells or tissues can be investigated using a variety of methods. Plant biologists use these techniques to gain information of genes involved in changes during development, or in response to endogenous or exogenous stimuli. Once identified these genes can be isolated, analysed and recombined, to provide further evidence for their function, and used with the aim of causing, or preventing, specific desired traits in plants, for academic or commercial purposes.

Most of the techniques described in this volume for cloning novel, differentially expressed genes are of use in other biological fields. Here they are written by experienced workers from their particular knowledge as plant biologists. This book is intended for practising scientists that have some experience with basic molecular biology. It gives an update on some well-established methods, and presents reliable protocols for new techniques extending the possibilities. The detailed methods in Chapters 1 to 6, describing the identification and isolation of such genes, can be used as a bench manual. Chapter 7 gives an overview of strategies and methods for further gene analysis. In each chapter the scope and the limitations of the methods presented are discussed, examples and possible reasons for unexpected results are given, and references are provided. We thank all the authors for their contributions and for their care in compiling these practical accounts of their specialist areas. We hope that the readers will find this book a useful addition to standard molecular biology manuals.

Ekki Hansen and Glyn Harper

Abbreviations

2D-PAGE	two-dimensional polyacrylamide gel electrophoresis
A	absorbance
amp	ampicillin
APS	ammonium persulphate
bp	base pairs
BSA	bovine serum albumin
cDNA	complementary DNA
Ci	Curie
CIAP	calf intestinal alkaline phosphatase
DEPC	diethylpyrocarbonate
DMF	dimethylformamide
DMSO	dimethylsulphoxide
dNTP	deoxynucleotide
DTT	dithiothreitol
EDTA	ethylenediaminetetraacetic acid
EMS	ethyl methanesulphonate
IEF	isoelectric focusing
kb	kilobases
K_m	Michaelis constant
MCS	multiple cloning site
MMLV	Moloney murine leukaemia virus
MUG	4-methylumbelliferone glucuronide
NP40	nonidet-P40
OD	optical density
PCR	polymerase chain reaction
PEG	polyethylene glycol
pfu	plaque-forming units
phage	bacteriophage
PNK	polynucleotide kinase
poly[A]$^+$RNA	polyadenylated RNA
PVDF	polyvinylidene difluoride

RNase	ribonuclease
mRNA	messenger RNA
rRNA	ribosomal RNA
tRNA	transfer RNA
rpm	revolutions per minute
SDS	sodium dodecyl sulphate
SDS–PAGE	sodium dodecyl sulphate–polyacrylamide gel electrophoresis
SSC	standard saline citrate
TCA	trichloroacetic acid
TEMED	N,N,N′,N′-tetramethylethylenediamine
TFA	trifluoroacetic acid
T_m	melting temperature
Tris	tris (hydroxymethyl) aminomethane
UV	ultraviolet
Xgal	5-bromo-4-chloro-3-indolyl-D-galactopyranoside
X-Gluc	5-bromo-4-chloro-3-indoyl-β-D-glucopyranoside

Differential Screening

CHARLES P. SCUTT

1.1 Introduction

Genes may be differentially regulated, leading to different mRNA transcript levels: between different tissues; at different developmental stages; in response to various environmental stimuli; or between genetically dissimilar individuals, e.g. mutant in contrast to wild-type plants. Differential screening is a method of identifying cloned DNA sequences homologous to differentially regulated genes. Early examples of such genes include the small subunit of ribulose-1,5-bisphosphate carboxylase (Bedbrook *et al.*, 1980) and α-amylase (Rogers and Milliman, 1983) and, to date, several hundred plant genes have been identified.

A differential screen, based only on differential mRNA levels, has the advantage of requiring no specific biochemical information relating to the genes to be investigated. cDNA probes, derived from mRNA preparations of different tissue samples, are used to screen a library. Clones from the library which are found to give rise to differential hybridization signals between the two probes are selected as differentially up-regulated in one of the tissues from which the probes were derived. Libraries are usually cDNA libraries constructed using one of the cDNA preparations used as differential probes. However, any collection of clones, such as other cDNA libraries, genomic libraries, polymerase chain reaction (PCR) products and plasmid DNA preparations, may be differentially screened. Subtractive hybridization methods can be designed to enrich libraries in differentially regulated cDNAs. Screening may be combined with other methods, for example to identify genes with homology to previously characterized genes that are expressed in specific tissues. In this case, differential screens might be performed on subpopulations of cDNAs previously isolated by heterologous probing of cDNA libraries. Alternatively, primers designed from conserved regions of a previously characterized gene or family can be used in PCR amplification of genomic DNA. The resulting amplified products may be differentially screened with total cDNA probes in order to identify differentially expressed members of that particular gene family.

1.1.1 Differential screening strategies

1.1.1.1 mRNA abundance: an important factor in differential screening

Many genes are differentially expressed and may be amenable to identification by differential screening. The probability of identification is dependent on the abundance of its transcript in the tissues selected for mRNA preparation. An efficient differential screening strategy should identify all classes of highly expressed, differentially regulated genes. Differential transcripts of high and intermediate expression levels may be identified by the differential screening of λ plaque-lifts or plasmid libraries in lysed bacterial colonies. However, differentially regulated genes expressed at low levels, e.g. typically those encoding transcription factors, signal transduction components, membrane receptors, etc., have a much lower chance of identification, unless combined with some other method. These transcripts may be identified through the screening of purified DNA preparations, 'cold-plaque' screening (Hodge *et al.*, 1992), or by using probes which have been enriched in differentially expressed cDNAs through subtractive hybridization procedures. Various strategies for the differential screening of cDNA libraries are illustrated in Figure 1.1.

1.1.1.2 Selection of plant tissues for differential screening

As described above, differential screening of λ phage plaques with unsubtracted cDNA cannot detect very low-abundance differential transcripts (< 0.02% occurrence in phage libraries, Hodge *et al.*, 1992) and poorly detects those of slightly higher abundance (0.02–0.05%, Sargent and Dawid, 1983). The abundance of differential transcripts in hybridization probes may be maximized by preparing these only from the particular tissues and at temporal stages from which maximal expression of the desired genes is expected. Many genes may be differentially regulated between two plant tissues – perhaps not all of principal interest to a particular investigation. In order, as far as possible, to limit the results of a screen to 'relevant' transcripts, the tissues from which mRNA samples are prepared for hybridization probe construction must be as closely matched as possible in all respects other than the differences under investigation.

1.1.1.3 Construction of cDNA libraries

cDNA libraries are usually constructed in λ addition vectors, rather than plasmid vectors, owing to ease of handling, storage, and high cloning efficiency. A method is given in this chapter (Protocols 1.3 to 1.8) for the synthesis of cDNA, which may be ligated into any cloning vector that includes an *Eco*RI cloning site. The cloning vectors in the λ Zap® series (Short *et al.*, 1988) are recommended as they allow great flexibility in library screening and also simplify the manipulation of cloned cDNAs. λ Zap® libraries facilitate the *in vivo* excision of recombinant phagemids, which enables the rapid recovery of cDNAs without subcloning. Kits for cDNA synthesis and library construction are available commercially and the use of these may be advantageous, particularly if few libraries are to be made. Additionally, a number of ready-made cDNA libraries are obtainable from academic collections or commercial companies.

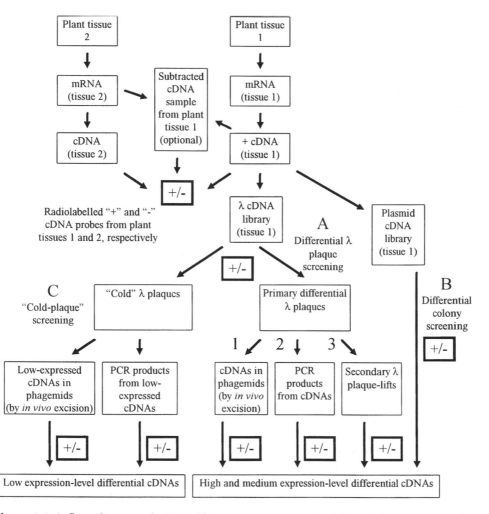

Figure 1.1 A flow diagram of cDNA library construction and differential screening in the cloning of cDNAs specifically expressed in one plant tissue (tissue 1), but not in another (tissue 2). Letters A, B and C indicate routes for bacteriophage λ plaque, bacterial colony and 'cold-plaque' differential screening, respectively. Numbers 1, 2 and 3 indicate three options for secondary differential screening of λ libraries. Boxed '+/−' symbols represent cDNA probes from plant tissues 1 and 2, and the use of these in various differential screening steps.

1.1.1.4 Differential screening of DNA preparations and 'cold-plaque' screening

Differential screening for low-abundance transcripts may be performed on plasmid DNA or PCR products. Screening of such DNA samples is more sensitive than plaque or colony screening due to the very low level of background and a correspondingly high signal-to-noise ratio. The recombinant plasmid DNA can be prepared from plasmid cDNA libraries, or following *in vivo* mass excision of λ Zap® cDNA libraries (see Protocol 1.12). PCR can be used to amplify cDNA inserts from recombinant λ phage plaques using λ cloning-site primer sequences (Protocol 1.13). Samples are immobilized on duplicate blots and differentially hybridized with labelled cDNA probes.

A modification on the differential screening of DNA preparations, termed *cold-plaque screening* (Hodge *et al.*, 1992), eliminates relatively abundant transcripts, typically 75% of an mRNA sample, by first selecting plaques which fail to give signals with either probe in a primary differential screen. Remaining 'cold' plaques, representing low-abundance transcripts, are further differentially screened on plasmid DNA dot-blots.

1.1.1.5 The use of subtracted probes in differential screening

Low-abundance cDNAs in a 'target' population may be identified by screening using probes which have been enriched by the elimination from the target population of cDNAs common to other 'driver' population(s). Enrichment of target cDNA probes is accomplished by subtractive hybridizations with 'driver' cDNA from tissues lacking expression of the differentially expressed genes. Subtractive hybridization may not completely remove abundant, constitutive transcripts from target cDNA samples and it is, therefore, worthwhile differentially screening cDNA libraries with both subtracted target probes and driver probes for comparison. Clones which give a strong signal only with subtracted positive probes are selected for further analysis. Subtraction hybridization is treated in detail, including protocols for the preparation of subtracted probes in Chapter 2.

1.1.1.6 Selection of probe-labelling method

Radioactive detection is a highly sensitive and reliable method for differential screening. Of the useful radioisotopes ^{32}P, ^{33}P and ^{35}S, labelling of probes with ^{32}P gives the greatest sensitivity of detection, able to detect as little as 10 fg of DNA (Evans and Read, 1992). A method based on labelling two differential hybridization probes with the isotopes ^{32}P and ^{35}S, respectively, has also been developed (Olszewski *et al.*, 1989). This method allows differential screening on a single set of plaque-lift filters and is described later in this chapter. Methods for non-radioisotopic detection based on enhanced chemiluminescence (ECL) are now available which claim to rival the sensitivity of ^{32}P-based methods for certain procedures (e.g. ECL, Amersham; RAD-FREE®, Schleicher & Schuell). These methods may be applied successfully to differential screening, particularly in laboratories in which they are already in routine use. However, for work requiring autoradiograph exposures of ≥4 days with ^{32}P, current ECL methods tend to compare unfavourably with radioisotopic methods, suffering from lower sensitivity and higher levels of background.

1.2 Preparation of RNA from plant tissues

1.2.1 The ribonuclease menace!

RNA is readily degraded by ubiquitous ribonucleases (RNases) and the human skin surface is a rich source of these. RNases are remarkably resistant to degradation, renaturing after autoclaving to regain their activity. For this reason, the following precautions should be taken when working with RNA:

- Wear latex gloves at all times when handling RNA and all solutions and equipment for RNA work.
- Bake glassware and spatulas for RNA work at 180°C for >2 h prior to use.
- Treat glass centrifuge tubes and solutions by overnight incubation at 37°C with 0.1% (v/v) of the ribonuclease inhibitor, diethylpyrocarbonate (DEPC).
- Autoclave solutions and glassware following incubation to destroy residual DEPC, to prevent RNA damage.

NB. Tris solutions cannot be treated with DEPC and should be made from RNase-free stocks using DEPC-treated sterile water.

- Keep separate stocks of chemicals and disposable plasticware for RNA work.

1.2.2 RNA extraction

The method for RNA extraction in Protocol 1.1 has been used successfully to prepare RNA from root, stem, leaf and flower tissues of species including *Arabidopsis thaliana*, *Nicotiana tabacum*, *Silene latifolia* and *Brassica oleracea*. This procedure involves the use of a lithium salt to precipitate RNA and despite the fact that lithium is an inhibitor of reverse transcriptase, no problem with carry-over of lithium ions into cDNA synthesis has been previously encountered following this extraction procedure. Alternative methods for RNA preparation, which may prove suitable for problematic plant species, are given by Croy *et al.* (1993). Commercial kits are widely available for the direct purification of mRNA from tissue samples.

PROTOCOL 1.1 EXTRACTION OF RNA FROM PLANT TISSUES

1. Grind frozen plant tissue to a fine powder using a mortar and pestle, pre-cooled with liquid nitrogen.
2. Add the tissue powder to a centrifuge tube[1] containing extraction buffer (1 ml/g of tissue) and phenol/chloroform (1 ml/g of tissue), held at 80°C in a water bath. Carefully disperse the tissue powder in the emulsion and place the sample on a rotary shaker to mix phenolic and aqueous phases for 30 min at room temperature.
3. Centrifuge the sample at 4000 rpm for 10 min, and remove the aqueous (upper) phase to a fresh centrifuge tube with an equal volume of chloroform, vortex and re-centrifuge.
4. Remove the aqueous (upper) phase to a fresh centrifuge tube and add 0.3 volume of 8 M LiCl, mix and incubate overnight at 4°C to precipitate RNA.
5. Centrifuge to recover RNA (8000 rpm, 30 min, 4°C) and discard the supernatant.
6. Wash the RNA-containing pellet in ice-cold 2 M LiCl (1 ml/g of starting material), centrifuge (10 min, 8000 rpm, 4°C) to recover the pellet.
7. Repeat step 6.
8. Dissolve the pellet in 1 ml/g starting material of 2% potassium acetate pH 5.5 and centrifuge (8000 rpm, 10 min, 4°C).

9. Remove the supernatant (containing RNA) from any insoluble material to a fresh centrifuge tube, add 2.5 volume ethanol and incubate at $-20°C$ overnight.

10. Recover the RNA pellet by centrifugation (8000 rpm, 30 min, 4°C) and wash in ice-cold ethanol (1 ml/g starting material). Centrifuge (8000 rpm, 10 min, 4°C) and remove the supernatant.

11. Dry the pellet under vacuum, resuspend the RNA in DEPC-treated, sterile distilled water and store under liquid nitrogen or at $-70°C$.

Notes

NB. Phenol is extremely toxic and caustic. Face and hand protection should be worn in addition to normal laboratory protective clothing. Procedures involving phenol should be carried out in a fume hood.

[1]15 and 30 ml Corex® (DuPont) tubes are suitable for processing 0.4–2 g and 2–5 g of tissue, respectively. Larger samples can be processed in 50 ml disposable polypropylene centrifuge tubes for initial centrifugation at 4000 rpm, and subsequently in 50 ml Corex® tubes. Small samples (< 0.4 g) may be processed entirely in 1.5 ml polypropylene centrifuge tubes, using a microcentrifuge at 13 000 rpm for all centrifugation steps.

Expect a maximum yield of 1 mg RNA /g tissue.

1.2.3 *Determination of RNA yield and integrity*

RNA may be quantified by UV spectrophotometry at 260 nm and 280 nm. In a pure RNA solution, which has an A_{260}:A_{280} ratio of 2.0, an A_{260} reading of 1.0 corresponds to an RNA concentration of 40 µg/ml. Deviations from the expected ratio may be caused by contaminants including DNA, phenol, polysaccharides and proteins. A 'shoulder' at 280 nm on a spectrophotometric scan may be caused by residual protein in RNA preparations.

RNA quality may be assessed by electrophoresis of samples of approximately 5 µg RNA/track through agarose gels containing formaldehyde stained with ethidium bromide, or denatured in glyoxal/DMSO and stained with acridine orange (Sambrook *et al.*, 1989). RNA may be detected with greater sensitivity (0.2 µg/ track) following Northern transfer to nylon or nitrocellulose membranes, by staining with 0.04% methylene blue in 0.5 M sodium acetate pH 5.2 and destaining in water. Following electrophoresis and staining of total RNA samples, 28S and 18S ribosomal RNA (rRNA) bands should be clearly visible. Particularly in leaf RNA samples, bands of chloroplast rRNAs may also be apparent.

1.2.4 *Poly[A]⁺RNA selection*

Poly[A]$^+$RNA selection is necessary to enrich samples in mRNA prior to cDNA synthesis. cDNA synthesis from total RNA tends to result in a high degree of ribosomal transcript contamination and should be avoided unless RNA samples are in extremely limited supply. A single round of poly[A]$^+$RNA selection results in enrichment of mRNA from approximately 1% (in total RNA) to more than 50%

(in poly[A]$^+$RNA samples), the remainder being largely rRNA. First-strand cDNA synthesis, primed from the poly[A]$^+$ tails of mRNA, further reduces the proportion of ribosomal transcripts represented in the cDNA. A method for the batch processing of RNA samples on oligo(dT) cellulose for poly[A]$^+$RNA selection is given in Protocol 1.2. First-strand cDNA products from this RNA template do not produce a hybridization signal against Southern dot-blots of immobilized ribosomal DNA. An alternative magnetic capture method of poly[A]$^+$RNA selection is described in Chapter 2.

It may be difficult to measure poly[A]$^+$RNA concentrations accurately if only small quantities are available. A reasonable estimate of poly[A]$^+$RNA concentration is 1% of the total RNA used in its preparation. The amount of cDNA synthesized for library construction may be accurately measured in the next stage of the procedure, from radioisotopic incorporation data.

PROTOCOL 1.2 POLY[A]$^+$RNA AFFINITY CHROMATOGRAPHY

1. Add oligo(dT) cellulose to 0.1 g/mg total RNA, in a sterile centrifuge tube (use 1.5 ml disposable tubes for ≤ 0.1 g oligo(dT) cellulose).

2. Wash oligo(dT) cellulose 4 times at 10 ml/g in 1× loading buffer, mix briefly and centrifuge at 2000 g, (or 6500 rpm in a microcentrifuge), for 2 min after each wash to pellet the oligo(dT) cellulose.

3. Dilute total RNA in DEPC-treated sterile water to a volume of 2 ml/g of oligo(dT) cellulose to be used. Incubate the RNA at 65°C for 5 min to remove any secondary structure, cool on ice, add 1 volume of 2× loading buffer and add the resulting solution to the washed oligo(dT) cellulose. Mix by gentle shaking for 5 min.

4. Centrifuge at 2000 g for 2 min and remove the supernatant (keep this for analysis, if required). Wash the oligo(dT) cellulose 5 times in 1× loading buffer (10 ml/g oligo(dT) cellulose, each wash), centrifuging (2000 g, 2 min) after each wash.

5. Elute poly[A]$^+$RNA by 4 sequential washes (centrifuge at 2000 g for 2 min after each wash) with elution buffer (2.5 ml/g oligo(dT) cellulose) and pool the eluates.

6. Precipitate, the poly[A]$^+$RNA by addition of 0.1 volume of 3 M sodium acetate pH 5.2 per 2.5 volume of ethanol. Incubate at −20°C for > 4 hours. Pellet the poly[A]$^+$RNA by centrifugation at 6000 g for 30 min. Wash the pellet in ice-cold 80% ethanol and dry under vacuum. Resuspend the poly[A]$^+$RNA in DEPC-treated, sterile distilled water and store it at −70°C or under liquid nitrogen.

To regenerate the oligo(dT) cellulose for reuse:

7. Wash the oligo(dT) cellulose 3 times (mix and centrifuge, as in step 2) in 0.1 M sodium hydroxide (10 ml/g oligo(dT) cellulose).

8. Wash sequentially in sterile water until the pH is near neutral (check eluate with pH paper).

9. Wash the oligo(dT) cellulose in ethanol (10 ml/g), desiccate and store at −20°C.

Notes

[1]If a low concentration (< 50 ng/ml) of RNA is to be precipitated, add 10 g glycogen (molecular biology grade) as a coprecipitant from a concentrated solution (10 µg/µl) and mix, prior to addition of ethanol.

1.3 cDNA synthesis and library construction

First-strand cDNA is synthesized on poly[A]$^+$mRNA templates by the action of reverse transcriptase, synthesis of cDNA being primed from a synthetic oligo(dT) which anneals to the poly[A]$^+$ tails of mRNA molecules. For library construction, cDNA must be rendered double-stranded, ligated to synthetic linkers and into a cloning vector. For cDNA probes, first-strand cDNA may be purified from unincorporated nucleotides by gel filtration (Protocol 1.8) and labelled by random priming (Protocol 1.10). First-strand cDNA for cloning is made double-stranded by replacement of the RNA strand of the mRNA–cDNA heteroduplex. DNA synthesis for this reaction is catalysed by *E. coli* DNA polymerase I, primed from nicks created by the action of RNAase H (Gubler and Hoffman, 1983). The 5'-end of the double-stranded cDNA is then polished by the action of T4 DNA polymerase. Ligation of cDNA into a cloning vector is most efficiently achieved following the addition of synthetic DNA linkers. The method for this given in Protocol 1.6 utilizes an adaptor which recreates an *Eco*RI half-site and can be ligated into *Eco*RI-cleaved, dephosphorylated λ or plasmid cloning vectors. The *Eco*RI-blunt end adaptor is constructed from one 5'-phosphorylated and one 5'-dephosphorylated oligonucleotide such that it may dimerize or ligate to cDNA, but cannot form concatamers. In the method given, no *Eco*RI methylation and restriction endonuclease treatment of the adaptored cDNA is necessary prior to ligation into a cloning vector, as is the case with some methods.

In Protocol 1.3 and certain commercial systems, the use of relatively large amounts of about 5 μg of poly[A]$^+$RNA for library construction is suggested. Such starting quantities enable simple monitoring of the whole process and provide excess material at some steps, should it be necessary to repeat any part of the process. However, if necessary, libraries of 10^6 recombinant phage can be constructed in λ cloning vectors using as little as 200 ng of poly[A]$^+$RNA starting material. Detailed protocols and information for handling λ libraries including vector/host strain genotypes, vector restriction maps, packaging extract preparation, λ DNA preparation and storage, and handling of libraries and phage stocks, are available in various reference texts (Huyhn *et al.*, 1985; Sambrook *et al.*, 1989; Brown, 1991).

1.3.1 Determination of cDNA yield and integrity

The course of the first- and second-strand cDNA synthesis reactions can be followed by incorporation of radioactive label in aliquots of pilot or main reactions. The label is detected by gel analysis and by scintillation counting both directly and following TCA precipitation of isotope-incorporated DNA (Sambrook *et al.*, 1989). The proportion of ^{32}P incorporated in each synthesis reaction, multiplied by the mass of dNTPs present (66 μg, Protocol 1.3) gives a value for the mass of cDNA produced (Equation 1.1).

$$\text{Mass of cDNA strand (μg)} = \text{activity of TCA precipitate(cpm)}/$$
$$\text{total activity(cpm)} \times 66(\text{μg}) \qquad (1.1)$$

PROTOCOL 1.3   FIRST-STRAND cDNA SYNTHESIS

1. Dilute ≤ 5 µg poly[A]$^+$RNA in DEPC-treated, sterile distilled water to a volume of 32 µl, incubate at 65°C for 5 min and chill on ice.

2. Add to the RNA the following reaction components:

    2.5 µl 1 M Tris-Cl pH 7.6

    3.5 µl 1 M KCl

    2.0 µl 250 mM MgCl$_2$

    1.0 µl human placental RNase inhibitor (40 units/µl)

    5.0 µl 0.1 M DTT

    2.0 µl 25 mM each of 4 dNTPs mix

    1.0 µl oligo(dT)15–18 (1 µg/µl)

3. Incubate for 10 min at room temperature to allow annealing of the oligo(dT) primer.

4. Add 1.0 µl of MMLV reverse transcriptase (100 units/µl).

5. Remove 5.0 µl of the reaction to a separate tube to which has been added 0.5 µl of one [α-^{32}P]dNTP as a pilot reaction to monitor the quality of first-strand cDNA. Omit this step if cDNA is to be used only for probe preparation.

6. Incubate main and pilot reactions for 1 h at 37°C.

For cDNA probe preparation, make the reaction up to 60 µl in 1× STE and purify first-strand cDNA by Sephacryl S400 gel filtration (Protocol 1.8).

PROTOCOL 1.4   SECOND-STRAND SYNTHESIS

1. To the main cDNA synthesis reaction add:

    2.8 µl 250 mM MgCl$_2$

    5.0 µl 2 M Tris-Cl pH 7.4

    1.5 µl 1 M (NH$_4$)$_2$SO$_4$

    87 µl distilled H$_2$O

    2.0 µl one [α-^{32}P] dNTP (10 µCi/µl)

    2.0 µl RNase H (1 units/µl)

    5.0 µl *E. coli* DNA polymerase I (10 units/µl).

2. Mix, pulse spin and incubate the reaction at 16°C for 2.5 h.

3. Remove three 1 µl aliquots of the second-strand reaction for scintillation counting and gel analysis.

4. Extract the second-strand reaction twice with equal volume of phenol/chloroform.

5. Add 0.1 volume 3 M sodium acetate pH 5.2 per 2.5 volume ethanol. Incubate for > 2 h at −20°C and recover the cDNA by centrifugation in a microcentrifuge for 30 min at 13 000 rpm.

6. Wash the pellet in 1 ml of ice-cold 80% ethanol, centrifuge (13 000 rpm, 2 min) and dry the pellet under vacuum.

PROTOCOL 1.5   POLISHING 5'-TERMINI OF cDNA

1. Resuspend the cDNA in 39.5 µl distilled water and add:

    5.0 µl T4 polymerase buffer

    2.5 µl 2.5 mM dNTP mix

    3.0 µl T4 DNA polymerase (3 units/µl)

2. Incubate at 37°C for 30 min.

3. Add 50 µl water to bring the total volume to 100 µl and phenol/chloroform extract the solution twice.

4. To the aqueous phase, add 0.1 volume sodium acetate per 2.5 volume ethanol and incubate at −20°C for ≤ 2 h.

5. Recover the cDNA by centrifugation at 13 000 rpm for 30 min, wash in ice-cold 80% ethanol, re-centrifuge (13 000 rpm, 2 min) and vacuum desiccate.

PROTOCOL 1.6   ADDITION OF *Eco*RI ADAPTORS

1. Anneal two oligonucleotides to give an *Eco*RI-blunt adaptor with a blunt-end 5' phosphate by mixing equal volume of 0.2 µg/µl oligos[1] A and B in 1× TE buffer.

2. Incubate the oligo mix briefly at 70°C, pulse-spin and allow to cool to room temperature.

3. Resuspend double-stranded cDNA for adaptor ligation in 8 µl of annealed adaptor and add:

    1.0 µl 10× kinase/ligase buffer

    1.0 µl T4 DNA ligase (5 units/µl)

4. Incubate at 16°C overnight.

5. Heat to 70°C for 30 min to inactivate the ligase.

Notes

[1]Oligo A, 5'-dephosphorylated is 5'(AATTCCCGGG); oligo B, 5'-phosphorylated is 5'(pCCCGGG), both available from New England Biolabs.

PROTOCOL 1.7   PHOSPHORYLATE 5'TERMINI

1. Add:

    1.0 µl 10x kinase/ligase buffer

    8.0 µl distilled water

    1.0 µl T4 polynucleotide kinase (10 units/µl)

2. Incubate the reaction at 37°C for 30 min.

3. Heat to 70°C for 15 min to inactivate the kinase.

PROTOCOL 1.8   PURIFICATION OF ADAPTOR-LIGATED cDNA

1. Mix a slurry of Sephacryl S-400 and pack into a 1 ml syringe barrel plugged with sterile siliconized glass wool to form a spin-column.

2. Place the spin-column in a 12 ml centrifuge tube (Falcon 2051) and centrifuge at 400 g (1500 rpm on a bench-top centrifuge) for 2 min.

3. Add further Sephacryl to 5 mm below top of the syringe, and re-centrifuge.

4. Wash the spin-column 5 times, each with 300 µl of 1× STE, (re-centrifuge after each addition).

5. Discard the wash solution from the Falcon tube and place a 1.5 ml microcentrifuge tube, without a cap, inside it. Replace the spin-column.

6. Adjust the adaptor–cDNA volume to 60 µl with 1× STE, add the cDNA sample to the spin-column and centrifuge as before, collecting the excluded fraction in the 1.5 ml tube.

7. Add 60 µl 1× STE to the column, centrifuge as before, add a further 60 µl of 1× STE and re-spin the column. The first three eluted fractions, now pooled beneath the spin column, contain cDNA of > 200 bp in length, and short cDNA fragments, adaptors and unincorporated nucleotides have been removed.

8. Ethanol precipitate the cDNA for > 3 h at −20°C following addition of 0.1 volume 3 M sodium acetate pH 5.2 per 2.5 volume of ethanol. Recover by centrifugation at 13 000 rpm for 30 min. Wash the pellet in 80% ice-cold ethanol, re-centrifuge (13 000 rpm, 2 min), vacuum desiccate and resuspend in 5 µl distilled water.

Notes

This protocol is also suitable for the purification of first-strand cDNA samples prior to random prime labelling (Protocol 1.10).

Some losses of yield will be encountered in steps subsequent to second-strand synthesis. If required, final yield of cDNA can also be calculated following scintillation counting of a small aliquot (e.g. 0.5 µl) of adaptor–cDNA prior to vector ligation, as cDNA has the same specific activity as the nucleotide solution from which it was made: 6.73×10^5 cpm/µg for Protocol 1.3. Equation 1.2 gives the mass of double-stranded cDNA in the aliquot taken for scintillation counting, but applies only on the activity date of the label: if one ^{32}P half-life (14.3 days) has expired on the date of scintillation counting, for example, then the final figure for yield of DNA would be twice that calculated from Equation 1.2.

$$\text{Mass of double-stranded cDNA}(\mu g) = \text{activity}(\text{cpm})/ \text{specific activity}(\text{cpm}/\mu g) \times 2 \qquad (1.2)$$

where specific activity $= 6.73 \times 10^5$ cpm/µg.

1.3.2 *Ligation of cDNA into cloning vectors and cDNA library construction*

For cloning in λ vectors, *Eco*RI–adaptor–cDNA is ligated into vector arms which have been cleaved with *Eco*RI and dephosphorylated with CIAP. Pre-cut, dephosphorylated vectors can be obtained commercially. Ligations should be performed overnight or longer at 15°C, with mass ratios of λ DNA to cDNA of approximately 2.5:1. If necessary, a series of ligations at different ratios may be tested. Prepared λ arms (ca. 0.5 μg) should be ligated with adaptor–cDNA in 1× kinase/ligase buffer using 1 μl T4 DNA ligase in a total volume of 5–10 μl.

Lambda ligations are packaged *in vitro* using protocols according to Sambrook *et al.* (1989) or high-efficiency commercial packaging extracts, e.g. Gigapack Gold®packaging extracts (Stratagene), to form primary libraries. Dilution series should be plated-out to determine primary library titres. Primary libraries may be plated directly for phage-lifting and differential screening, but it is usual to amplify libraries (Sambrook *et al.*, 1989), as amplified libraries are more stable in storage and provide sufficient phage for an unlimited number of experiments. Primary libraries in the λ Zap®II vector, containing unmethylated cDNA from Protocols 1.3–1.8, may be amplified in either XL1-blue or XL1-blue MRF′(Stratagene) *E. coli* strains. Primary λ Zap®II libraries constructed from cDNA made using methylated deoxynucleotides according to Stratagene UniZap XR protocols, must be amplified in XL1-blue MRF′ or SURE™ strains as these do not destroy such hemimethylated DNA. Amplified λ Zap®II libraries may be plated for plaque-lifts using either XL1-blue or XL1-blue MRF′ strains.

For small plasmid libraries, cDNA is ligated into suitably digested, CIAP-treated, plasmid vectors. Competent cells are transformed chemically (Hanahan, 1985), or by high-efficiency electroporation, as described in Chapter 2.

1.4 Primary differential screening of cDNA libraries

1.4.1 *Plating phage cDNA libraries for differential screening*

Differential screening of λ cDNA libraries requires phage to be plated (Protocol 1.9) at a lower density than does screening with a single-sequence probe, to enable visualization of hybridization signals from all individual plaques and to minimize the number of rounds of differential screening required before recombinants are obtained in pure form. Plaque density for differential screening should be approximately 5000 pfu/14-cm diameter plate (30 pfu/cm^2). Individual plaques which show differential signals may be identified following first-round differential screening and cored from agar plates. However, a second round of screening is usually required to verify that differential cDNAs from primary screens are genuine and to recover them in a purer form.

Transcripts of less than 0.02% abundance fail to give signals in differential plaque hybridization (Hodge *et al.*, 1992). A screened primary library of 25 000 plaques has a 99% probability of occurrence at least once, and a mean occurrence of five plaques, for the least abundant clone which can be detected by differential screening. Consequently, there is no merit in screening a library larger than this and smaller libraries will suffice in many cases. At a plating density of 30 plaques/cm^2, a 25 000 plaque library requires five 14 cm diameter or two 20 × 20 cm plates.

PROTOCOL 1.9   PHAGE λ PLATING AND PLAQUE LIFTING

1.   Inoculate 50 ml of L broth containing 0.2% (w/v) maltose in a 250 ml conical flask, with 0.5 ml of an overnight culture of an appropriate *E. coli* plating strain for the λ vector in

use. Grow the plating culture at 37°C, with shaking to A_{600} = 0.1–0.5, for approximately 2–4 h.

2. Pre-warm NZYM agar plates (14 cm diameter or 20 × 20 cm) at 37°C for 1 h.

3. Spin down the plating culture at 2000 rpm in a bench-top centrifuge for 10 min and gently resuspend in 10 mM $MgSO_4$ to A_{600} = 0.5.

4. Mix 600 μl plating cells and 5000 pfu λ library for each 14 cm plate (or 1.5 ml cells and 12 500 pfu for 20 × 20 cm plates), in sterile tubes with caps and incubate these at 37°C for 15 min to allow phage adhesion.

5. Add 8 ml NZYM top agarose if using 14 cm plates, or 20 ml if using 20 × 20 cm plates, from a molten stock held at 45°C to the infected cultures. Mix the tube contents together by single inversion, pour onto prewarmed NZYM plates and spread the top agarose by circular movement of plates on the bench-top such that the top agarose coverage becomes uniform. Leave for 5 min at room temperature for the agarose to gel.

6. Incubate the plates, inverted at 37°C, until plaques are about 1 mm diameter (between 5 and 10 h, depending on strain), remove the plates and chill at 4°C for > 2 h or overnight if convenient, before further processing.

7. Lower nylon[1] (or nitrocellulose) membranes onto the plaque-lift plates – centre first, holding with gloved hands at the membrane edges. Use an ink-filled syringe and wide-bore needle to create three registration marks per plate by stabbing through the membrane into the agar. Allow the first filter of each duplicate pair to transfer for 1–2 min and the second filter to transfer for 2–4 min. Create registration holes in the second of each filter pair, identically aligned with those of the first filter, using the ink-filled syringe and needle.

8. Incubate the plaque-lift filters sequentially in dishes of denaturing solution and neutralizing solution for 5 min each, plaque-side up, the filters may be immersed. Transfer the filters to blotting paper to dry.

9. Wrap nylon filters in Saranwrap and expose them on a UV transilluminator or light-box (e.g. Stratalinker™, Stratagene) to cross-link DNA to the filters. Times of exposure on transilluminators should be previously calibrated for the UV source. If nitrocellulose filters have been used, these should be baked at 80°C for 1 h.

Notes

[1]Nylon membranes such as Hybond-N (Amersham) are much more robust in use than nitrocellulose and no great difference in sensitivity or background exists between these two membrane types.

1.4.2 *Labelling of probes for differential screening*

Random primed labelling (Feinberg and Vogelstein, 1984), is the method of choice for the radiolabelling of cDNA probes for use in differential screening. A simplified version of this labelling technique, applicable to both first-strand and double-stranded cDNAs after purification from residual unincorporated nucleotides from cDNA synthesis, is given in Protocol 1.10. It is not advisable to label cDNA probes directly during first-strand cDNA synthesis as this is less economical in the use of valuable mRNA samples and because reverse transcriptase has a lower substrate affinity for deoxynucleotide substrates (K_m = 31 μM, dCTP; 24 μM, dATP) than

Klenow polymerase (K_m = 2 µM, mean value). Deoxynucleotide label of high (3000 Ci/mmol) or standard (400 Ci/mmol) specific activity may be used for random prime labelling, though the latter will be a less economical use of cDNA as it requires more template to achieve a high percentage of label incorporation. In using 50 µCi of label of high specific activity, incorporation of 60–80% should be achievable with 25 ng of cDNA template, whereas label of 400 Ci/mmol specific activity may require up to 250 ng of cDNA for equivalent radioisotopic incorporation.

PROTOCOL 1.10   cDNA PROBE LABELLING BY RANDOM PRIMING

1. Mix first-strand cDNA[1], 10 µl of random-sequence, hexameric DNA (18 OD units/ml) and water to a total volume of 33 µl.

2. Heat to 100°C for 5 min and pulse-spin.

3. Add to the cDNA/primer mix:

   10 µl 5 × labelling buffer

   5 µl [α-^{32}P] dCTP or [α-^{32}P] dATP (10 µCi/µl, 400–3000 Ci/mmol)

   2 µl Klenow polymerase (2 units/µl)

4. Mix, pulse-spin and incubate at 37°C for 2 h.

5. Separate the probe from unincorporated nucleotides by gel-filtration through Sephadex G-50 or an equivalent matrix in a spin column (Sambrook *et al.*, 1989), or NucTrap® push column (Stratagene) or similar.

6. Heat denature the probe (100°C,10 min), chill on ice and pulse spin to collect condensed liquid prior to addition to a hybridization solution.

Notes

[1]Use approximately 25 ng and 250 ng of cDNA template for dNTP labels of specific activities of 3000 Ci/mmol and 400 Ci/mmol, respectively.

1.4.3 *Differential plaque hybridization*

Primary differential screening may be performed either in sealed polythene bags and boxes in a shaking water bath or more conveniently in a hybridization oven (e.g. Hybaid or Techne). If hybridization ovens are used, multiple or overlapping filters should be separated from each other in each hybridization bottle using sheets of nylon mesh (30 µm pore size, available from John Stanier & Co., Manchester, UK). Duplicate sets of plaque-lift filters are prewashed several times to remove bacterial debris, prehybridized to block the membranes and hybridized to radiolabelled probes. Polyuridylic acid (250 µg/ml final concentration) can be added prior to hybridizations with cDNA probes made directly from mRNA templates, to reduce or eliminate non-specific hybridization of probe molecules to the poly[dA] tails of cDNA library clones. Similarly, for second-strand cDNA probes primed from unlabelled first-strand cDNA, polydeoxyadenylic acid can be used as a blocking agent. However, tracts of A:T hybrids have a low melting temperature and omission of nucleic acid homopolymers from prehybridization solutions seems to have no adverse effect on the results obtained. Removal of excess probe following filter hybridization is performed by sequential washes at low stringency and a final high-stringency wash. Where related gene family members are constitutively expressed then lack of high-stringency washing may lead to differential plaques remaining undetected.

PROTOCOL 1.11 DIFFERENTIAL PLAQUE HYBRIDIZATION

1. Prewash hybridization filters in large volumes of 3 × SSC/0.1% (w/v) SDS with shaking or rotation at 65°C for ≥2 h per wash until the wash solutions no longer appear discoloured or smell of bacterial debris.

2. Prehybridize the filters in prehybridization solution for ≥2 h at 65°C. Filter sets for each cDNA probe to be used may be prehybridized/hybridized together in the same polythene bag or bottle. Ensure there is a free flow of solution around all parts of the filters.

3. Add denatured, radiolabelled probes to the hybridizations and continue to incubate with shaking/rotation at 65°C for 16–24 h.

4. Following hybridization, wash the filters sequentially in 50–200 ml aliquots of pre-warmed 2 × SSC/0.1% (w/v) SDS at 65°C for 20 min per wash. Continue the washes until no extra probe can be removed, usually requiring 3–4 washes (monitor waste wash solutions with a Geiger counter). Finally wash the filters in 0.1 × SSC/0.1% (w/v) SDS at 65°C for 10 min.

5. Wrap the filters in Saranwrap, mount on card or old films, create radioactive ink registration marks and expose the filters to X-ray films, initially overnight, and subsequently for up to 2 weeks, as required.

1.4.4 *Examination of autoradiographs for differential cDNAs*

An initial overnight exposure to X-ray film is used to determine approximate signal strength and further exposure times of autoradiographs for positive and negative probes are adjusted such that non-differential plaques show approximately equal signals. Exposure times may be extended up to 2 weeks (one ^{32}P half-life). The autoradiographs are aligned with their respective plaque-lifts using radioactive ink marks and filter registration holes are marked in fine-line pen on the X-ray films. Duplicate autoradiographs are superimposed and scrutinized carefully over a light box for differential hybridization signals, which should be encircled on the positive autoradiograph. A rapid computer-aided method has also been developed for this process (Tizard *et al.*, 1994), which subtracts images to give a difference image from pairs of autoradiographs. False positives will be encountered in library screening; these invariably become apparent on further rounds of differential screening. Genuine plaque hybridization signals tend to be of slightly diffuse appearance, lacking 'hard' centres. Occasionally, these signals have short, comet-like tails derived from plaque lifting. Following identification of differential plaque locations, positive autoradiographs on which differential signals have been marked are aligned with phage plates. The autoradiographs should be placed on a light box and the plates superimposed such that registration marks are exactly aligned. Phage plugs from single plaques overlying differential signals are cored using sterile narrow Pasteur pipettes and expelled into 1.5 ml tubes containing 1 ml of 4% (v/v) chloroform in SM buffer, vortexed and stored at 4°C.

1.4.5 *Plating and differential screening of plasmid libraries*

The screening of plasmid libraries is no longer a common practice, having been largely superseded by techniques using advanced λ addition vectors. However, to its credit, the differential screening of lysed bacterial colonies yields pure differential cDNA clones, free from other contaminating clones, after only one screening step. A small library of bacterial clones may be stored at −80°C as separate glycerol stocks in the wells of microtitre plates, as described by Mason and Williams (1985). In this method, cDNA clones are plated-out using a specially constructed tool to replicate groups of 48 colonies (half of a microtitre plate) simultaneously onto duplicate sets of hybridization filters. The filters are incubated on antibiotic-containing agar plates to allow growth of bacterial colonies and then processed for nucleic acid screening by alkaline denaturation. This method for the storage of clones is very convenient and allows for many replications of libraries into fresh microtitre plates for storage or onto further filters for screening. The main disadvantage of this method is that it is labour intensive in the original picking of transformants and, therefore, not recommended for the storage of more than approximately 1000 cDNA clones.

1.5 Secondary differential screening of cDNA libraries

Secondary screening of primary differential cDNAs is necessary as primary differential signals may be false positives, often arising due to inconsistencies between 'duplicate' plaque-lift filters, and also because cross contamination of plaques may occur during plaque lifting or storage. Secondary screening involves re-screening putative first-round positive cDNAs, either as phage-lifts or DNA preparations, with freshly labelled cDNA probes. Reuse of labelled probes from primary differential screening is possible, but is generally not advisable as these tend to be depleted in the relevant hybridizing transcripts. There are essentially three options for secondary differential screening, numbered 1 to 3 in Figure 1.1, involving either *in vivo* excision, PCR screening or secondary phage-lift screening. The *in vivo* excision option is the most rapid and simple method, but can only be used for libraries in λ Zap® vectors or other vectors engineered for *in vivo* excision (e.g. Swaroop *et al.* 1991).

1.5.1 *Secondary screening following* in vivo *excision*

Libraries in λ Zap® II vectors can be excised as recombinant pBluescript™ II SK-phagemids *in vivo* from cored λ plaques. The *in vivo* excision technique is based on the property of pBluescript™ II phagemids, which have both Col E1 plasmid and f1 phage origins of replication, to be excised from λ Zap™ II vectors, following super-infection with the amber-mutated ExAssist™ (Stratagene) helper phage. The phagemid replicates as single-stranded DNA, is packaged and released from cells of the amber-suppressing (SupE) *E.coli* strain XL1-blue. Recombinant pBluescript™ II phagemid particles may then enter and replicate in host cells of the *E. coli* SOLR™ strain, but this non-suppressing host prevents further replication of the ExAssist™ helper phage. A modification of the *in vivo* excision method of Hay and Short (1992) is given in Protocol 1.12. The modified method facilitates the excision of large numbers of recombinant phagemid clones, and is accomplished in single 1.5 ml tubes. It yields tens to hundreds of bacterial transformants per original λ plaque. Plasmids should be isolated from two or three bacterial colonies recovered per original plaque from the *in vivo* excisions. The plasmid minipreparation method of Holmes and Quigley (1981), as adapted by Sambrook *et al.* (1989), is a suitable method for handling large numbers of samples. In order to confirm which excised cDNAs from primary screening are genuinely differentially expressed, recombinant plasmids containing cDNAs are immobilized on pairs of duplicate DNA dot-blots (Marzluff and Huang, 1984; or a similar method) and differentially screened with labelled cDNA probes (Protocol 1.10).

PROTOCOL 1.12 *IN VIVO* EXCISION OF RECOMBINANT λ ZAP®II

1. Inoculate XL1-blue and SOLR™ *E. coli* strains into universal bottles containing 10 ml L broth (supplemented with 10 mM MgSO$_4$ and 0.2% (w/v) maltose for XL1-blue) from 0.1 ml of an overnight culture. Grow the cultures to mid-log phase at 37°C with shaking, for ca. 2–4 h to reach A_{600}= 0.1–0.5.

2. Add 200 µl aliquots of XL1-blue culture to 1.5 ml tubes and to these add 5 µl λ phage solution from individual cored plaques in SM buffer (or 5 µl amplified λ Zap™II library for mass excision)[1] and 10^7 ExAssist M13 helper phage (1 µl from 10^{10} pfu/ml stock).

3. Incubate cultures at 37°C for 2 h with shaking (for excision) and 65°C for 15 min to kill XL1-blue cells, then cool to 37°C.

4. Add 200 µl aliquots of SOLR™ plating-cell culture to excision cultures from step 3, mix and incubate at 37°C for 15 min.

5. Spread 200 µl aliquots from excised cultures on L-ampicillin plates and incubate these overnight at 37°C to allow the growth of colonies.

Notes

[1] This method can be used in several tubes in parallel for mass excisions from whole λ libraries if less than approximately 1000 excised recombinant phagemids are required. More efficient mass excisions can be obtained using the more involved methods of Hay and Short (1992) or Owens *et al.* (1991). The latter utilizes the more efficient VCS-M13 helper phage, though growth of this in a second host cannot be prevented and problems of helper phage contamination in transformants may be encountered.

1.5.2 *Secondary screening of PCR products*

Secondary screening in vectors which do not permit *in vivo* excision is most conveniently achieved by a PCR-based method, in which primary differential cDNAs are amplified from cores of λ plaques using primers derived from λ vector sequences flanking the cDNA cloning site (Protocol 1.6). Similar PCR secondary differential screening procedures have also been described by Luo *et al.* (1994) and Thomas *et al.* (1994). PCR reactions are performed from each differential λ plaque core taken from primary plaque-lift plates, the PCR products are analysed on 1% agarose gels and bi-directionally Southern blotted onto duplicate hybridization membranes (Smith and Summers, 1980; also given in Sambrook *et al.*, 1989). These duplicate membranes are processed to denature the DNA and screened differentially using radiolabelled cDNA probes, as described for the differential screening of primary libraries (omitting the prewash, step 1). Following autoradiography, the positions of differential bands are accurately noted from DNA size markers and the corresponding DNA fragments are isolated from other agarose gels on which the same PCR products have been analysed. These differential cDNAs are then subcloned in plasmid cloning vectors.

PCR products obtained using *Taq* polymerases may contain a relatively high rate of mutations (ca. 1 in 400 bp), caused through replication errors. The fidelity of PCR products may be improved by the use of proof-reading polymerases such as Vent™, DeepVent™ (New England Biolabs) or *Pfu* (Stratagene). PCR products generated by the action of *Taq* polymerase, which has a terminal deoxyadenosyl transferase activity, should be cloned in modified vectors such as the TA vector (Invitrogen). PCR products obtained using Vent™ or *Pfu* polymerases may be treated with T4 polynucleotide kinase to phosphorylate 5′-termini and ligated into dephosphory-lated, blunt-end digested, plasmid cloning vectors. A list of suitable PCR amplification primers for use with various λ strains is given in Protocol 1.13 and further primer pairs may be designed for any other λ vector. More details on PCR protocols and the design of PCR primers may be found in Fordham-Skelton *et al.* (1993), Newton and Graham (1994), McPherson *et al.* (1995) and references therein.

PROTOCOL 1.13 PCR AMPLIFICATION OF RECOMBINANT λ PHAGE INSERTS

1. Core primary differential plaques from plates using the narrow ends of sterile Pasteur pipettes, and expel into 200 μl aliquots of SM buffer in 1.5 ml tubes and mix by vortexing.

2. Add 1 μl of phage suspensions from step 1, to 49 μl aliquots of PCR reaction mixes, to give final concentrations of:

   1 × *Taq*/Amplitaq™ buffer

   0.2 mM each of all 4 dNTPs

   1 μM each of 2 PCR primers[1]

   0.5–1 units *Taq*/Amplitaq™ polymerase

   Overlay the reactions with 50 μl of mineral oil, if necessary.

3. PCR thermal cycles are:

   [94°C × 30 s, 45°C[2] × 30 s, 72°C × 2 min] × 30

   [72°C × 10 min] × 1

4. Analyse the PCR products on 1% (w/v) agarose gels[3] with molecular size markers in the range 0.1–3 kb.

Notes

[1] Primer pairs which work for various λ vectors are as follows: for λgt11, 5′(TCAACAGCAACTGATGGAAACCAG) and 5′(TTGACACCAGACCAAC-

TGGTAATG) oligonucleotide primers; for λ GEM™2/ λ GEM™4, SP6 and T7 primers; for λ Zap®/ λ Zap® II/ λ Zap® Express, T3 and T7 primers.

[2] The temperature of the annealing step of the PCR reaction is primer-dependent: 45°C has been found to work well for T3, T7 and SP6 promoter primers (17–18-mers).

[3] Gels may be bi-directionally Southern blotted onto duplicate nylon or nitrocellulose membranes for secondary screening according to Sambrook *et al.* (1989).

1.5.3 *Secondary plaque-lift screening*

Secondary screening of phage-cloned cDNAs can be achieved by the differential screening of secondary plaque-lifts. Phage λ stocks from primary differential plaques should be titred and plated-out on separate plates, each at a density of approximately 50–250 pfu per 8 cm diameter plate. Plaque-lifts are processed and screened with radiolabelled cDNA probes as for primary differential screening. Selected differential plaques from each plate are removed and grown for λ DNA preparation (Grossberger, 1987). These λ DNA samples are digested to release cDNA inserts, which are then subcloned into plasmid vectors by standard methods (Sambrook *et al.*, 1989). The only advantage of this method of secondary screening, which is slow and labour intensive, is that cDNAs are never subjected to PCR amplification.

1.6 Dual-labelling with ^{32}P and ^{35}S in differential screening

The dual-labelling method (Olszewski *et al.*, 1989) is based on the use of two radio-isotopes of differing β-emission energies to label two different cDNA samples. In this method the target probe, containing the differential cDNAs of interest, is ^{35}S-labelled, whilst the reference probe is ^{32}P-labelled. Labelled cDNAs are used together to probe a single set of plaque-lift filters, dot-blots or any other array of samples to be differentially screened. Two exposures of the hybridization filters are made following washing: one fluorographic exposure and one autoradiographic exposure through a barrier which attenuates only the lower energy ^{35}S β-particles. Differential hybridization signals, occurring only on the X-ray film exposed without attenuation, result from ^{35}S emissions alone and represent transcripts which are specific to the target cDNA sample. The dual-labelling procedure requires only one set of hybridization membranes for differential screening and thereby eliminates a proportion of false positives which otherwise arise from differences between 'duplicate' plaque-lift filters.

Preparation of probes for dual-label hybridizations may be performed essentially according to Protocol 1.10, substituting [α-^{35}S]thio-dNTP for [α-^{32}P]dNTP in labelling target probes, and using similar amounts and specific activities of ^{32}P- and ^{35}S-labels for reference and target probes, respectively. cDNA probes should be heat-denatured separately prior to addition to the hybridization membranes. Autoradiography/fluorography of membranes, following washing, may be most sensitively carried out for both target and reference probes simultaneously. Membranes are sprayed with EnHance (Kodak) or a similar fluorographic spray and assembled with Kodak X-Omat AR or equivalent films, with double-sided emulsion, as indicated in Figure 1.2. A less sensitive alternative arrangement for autoradiography and

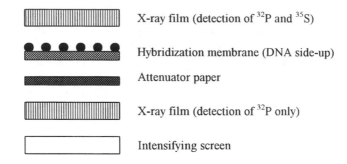

X-ray film (detection of ^{32}P and ^{35}S)

Hybridization membrane (DNA side-up)

Attenuator paper

X-ray film (detection of ^{32}P only)

Intensifying screen

Figure 1.2 The arrangement of filters, screens and X-ray films used for autoradiography/fluorography in dual-label differential screening (after Olszewski *et al.*, 1989).

fluorography is also given by Olszewski *et al.* (1989) for use if reference probe images are found to be too diffuse, due to the distance between filters and X-ray films. However, image resolution should not be a problem if the filter and film assemblies are tightly compressed together. The attenuator of ^{35}S β-electrons used by Olszewski *et al.* (1989) consists of yellow interleaf paper from X-ray film packaging, though other paper of similar opacity and weight can be substituted for this.

1.7 Solutions

Denaturing solution: 1.5 M NaCl, 0.5 M NaOH.

Elution buffer: 10 mM Tris-Cl pH 7.6, 1 mM EDTA, 0.05% (w/v) SDS.

Extraction buffer: 0.1 M LiCl, 1% (w/v) SDS, 0.1 M Tris-Cl pH 8.9.

Kinase/ligase buffer, 10×: 0.5 M Tris-Cl pH 7.6, 100 mM MgCl$_2$, 100 mM DTT, 500 µg/ml BSA (Fraction V, Sigma).

Labelling buffer, 5×: 0.25 M Tris-Cl pH 7.6, 50 mM MgCl$_2$, 100 µM each of three dNTPs and excluding the labelled dNTP in use.

L broth and other media for bacterial growth as described in Sambrook *et al.* (1989).

Loading buffer 2x: 40 mM Tris-Cl, pH 7.6, 1.0 M NaCl, 2 mM EDTA, 0.2% (w/v) sodium lauryl sarcosine.

Neutralising solution: 1.5 M NaCl, 0.5 M Tris-Cl pH 8.0.

Prehybridization solution: 6 × SSC, 5 × Denhardt's reagent, 0.1% (w/v) SDS, 0.05% (w/v) tetrasodium pyrophosphate, 40 µg/ml sheared, denatured non-homologous DNA (e.g. from herring testes or calf thymus). Prehybridization solution should be sterilized by filtration through a 0.2 µm pore filter prior to addition of the non-homologous DNA.

SM buffer: 5.8 g NaCl, 2.0 g MgSO$_4$, 50 ml 1M Tris-Cl pH 7.5, 5 ml 2% (w/v) gelatin per litre.

SSC, 20x: 3M NaCl, 0.3M trisodium citrate, pH 7.0 with HCl.

STE: 0.1 M NaCl, 20 mM Tris-Cl pH 7.5, 10 mM EDTA.

Taq buffer: 10 mM Tris-Cl pH 8.3, 50 mM KCl, 1.5 mM MgCl$_2$; 0.1% (w/v) gelatin.

TE buffer: 10mM Tris-Cl pH 7.6, 1mM EDTA.

T4 polymerase buffer, 10×: 0.33 M Tris-acetate pH 8.0, 0.66 M potassium acetate, 0.1 M magnesium acetate, 5 mM DTT, 1 mg/ml BSA (Fraction V, Sigma).

References

BEDBROOK, J.R., SMITH, S.M. and ELLIS, R.J., 1980, Molecular cloning and sequencing of cDNA encoding the precursor to the small subunit of chloroplast ribulose-1,5-bisphosphate carboxylase, *Nature*, **287**, 692–697.

BROWN, T., 1991, *Molecular Biology Labfax*, Oxford: BIOS Scientific Publishing Ltd.

CROY, E.J., IKEMURA, T., SHIRSAT, A. and CROY, R.R.D., 1993, Plant nucleic acids, in Croy, R.D.D. (Ed.) *Plant Molecular Biology Labfax*, pp. 21–48, Oxford: BIOS Scientific Publishing Ltd.

EVANS, M.R. and READ, C.A., 1992, ^{32}P, ^{33}P, and ^{35}S: selecting a label for nucleic acid analysis, *Nature*, **358**, 520–521.

FEINBERG, A.P. and VOGELSTEIN, B., 1984, A technique for radiolabeling DNA restriction fragments to high specific activity: addendum, *Analytical Biochemistry*, **137**, 266–267.

FORDHAM-SKELTON, A., EVANS, I.M. and CROY, R.R.D., 1993, Plant gene expression and PCR techniques, in Croy, R.D.D. (Ed.) *Plant Molecular Biology Labfax* pp. 313–372, Oxford: BIOS Scientific Publishing Ltd.

GROSSBERGER, D., 1987, Minipreps of DNA from bacteriophage lambda, *Nucleic Acids Research*, **15**, 6737.

GUBLER, U. and HOFFMAN, B.J., 1983, A simple and very efficient method for generating cDNA libraries, *Gene*, **25**, 263–269.

HANAHAN, D., 1985, Techniques for transformation of *E. coli* in Glover, D.M. (Ed.) *DNA Cloning: A Practical Approach*, Vol. 1., pp. 49–78, Oxford: IRL Press.

HAY, B. and SHORT, J.M., 1992, ExAssist™ helper phage and SOLR™ cells for λ Zap®II excision, *Strategies in Molecular Biology*, **5**, 16–18.

HODGE, R., PAUL, W., DRAPER, J. and SCOTT, R., 1992, Cold-plaque screening: a simple technique for the isolation of low abundance, differentially expressed transcripts from conventional cDNA libraries, *The Plant Journal*, **2**, 257–260.

HOLMES, D.S. and QUIGLEY, M., 1981, A rapid boiling method for the preparation of bacterial plasmids, *Analytical Biochemistry*, **114**, 193.

HUYNH, T.V., YOUNG, R.A. and DAVIS, R.W., 1985, Constructing and screening cDNA libraries in λgt10 and λgt11, in Glover, D.M. (Ed.) *DNA Cloning: A Practical Approach*, Vol. 1., pp, 49–78, Oxford: IRL Press.

LUO, G., AN, G. and WU, R., 1994, A PCR differential screening method for rapid isolation of clones from a cDNA library, *BioTechniques*, **16**, 672–675.

MARZLUFF, W.F. and HUANG, R.C.C., 1984, Transcription of RNA in isolated nuclei, in Hames, B.D. and Higgins, S.J. (Eds) *Transcription and Translation: A Practical Approach*, pp. 89–128, Oxford: Oxford University Press.

MASON, P.J. and WILLIAMS, J.G., 1985, Hybridisation in the analysis of recombinant RNA, in Hames, B.D. and Higgins, S.J. (Eds) *Nucleic Acid Hybridisation: A Practical Approach*, pp. 113–137, Oxford: IRL Press.

McPHERSON, M.J., HAMES, B.D. and TAYLOR, G.R., 1995, *PCR 2: A Practical Approach*, Oxford: Oxford University Press.

NEWTON, C.R. and GRAHAM, A., 1994, *PCR*, Oxford: BIOS Scientific Publishers Ltd.

OLSZEWSKI, N.E., GAST, R.T. and AUSUBEL, F.M., 1989, A dual-labeling method for identifying differentially expressed genes: use in the identification of cDNA clones that hybridise to RNAs whose abundance in tomato flowers is potentially regulated by gibberellins, *Gene*, **77**, 155–162.

OWENS, G.P., HAHN, W.E. and COHEN, J.J., 1991, Identification of mRNAs associated with programmed cell death in immature thymocytes, *Molecular and Cellular Biology*, **11**, 4177–4188.

ROGERS, J.C. and MILLIMAN, C., 1983, Isolation and sequence analysis of a barley α-amylase cDNA clone, *Journal of Biological Chemistry*, **258**, 8169–8174.

SAMBROOK, J., FRITSCH, E.F. and MANIATIS, T., 1989, *Molecular Cloning: A Laboratory Manual*, 2nd Edn, USA: Cold Spring Harbor Laboratory Press.

SARGENT, T.D. and DAWID, I.B., 1983, Differential gene expression in the gastrula of *Xenopus laevis*, *Science*, **222**, 135–139.

SHORT J.M., FERNANDEZ, J.M., SORGE, J.A. and HUSE, W.D., 1988, Lambda-Zap: a bacterio-phage expression vector with *in vivo* excision properties. *Nucleic Acids Research*, **16**, 7583–7600.

SMITH, G.E. and SUMMERS, M.D., 1980, The bidirectional transfer of DNA and RNA to nitrocellulose or diazobenzyloxymethyl paper, *Analytical Biochemistry*, **109**, 123.

SWAROOP, A., XU, J., AGARWAL, N., and WEISSMAN, S. M., 1991, A simple and efficient cDNA library subtraction procedure: isolation of human retina-specific cDNA clones, *Nucleic Acids Research*, **19**, 1954.

THOMAS, M.G., HESSE, S.A., AL-MAHDAWI, S., BUI, T.D., MONIZ, C.F. and FARZANAH, F., 1994, A procedure for second-round differential screening of cDNA libraries, *BioTechniques*, **16**, 229–232.

TIZARD, B.T., STANTON, J.-A.L. and LAING, N.G., 1994, DiffScreen: the merging of image subtraction and molecular genetics for the rapid analysis of differentially screened cDNA libraries, *CABIOS*, **10**, 209–210.

Subtractive cDNA Cloning

GLYN HARPER

2.1 Introduction

Subtractive hybridization is a powerful method of detecting and isolating gene sequences that are differentially expressed. Conventionally, differential hybridization has been used to detect such sequences, and procedures are described in this volume. Practically, however, it cannot be used to detect sequences with an abundance lower than approximately 0.1% of the original population (Sambrook *et al.*, 1989: Hodge *et al.*, 1992). Many transcripts in eukaryotic cells are well below this level of abundance and so cannot be detected by this method. Suitable methods, described in this volume, include cold-plaque screening, differential display and subtractive hybridization.

In subtractive hybridization, sequences common to two or more populations are subtracted from one of the populations, giving an effective enrichment for unique or up-regulated sequences. These sequences are separated from the rest, for cloning, analysis or use as probes. The terminology commonly describes the population of interest as the [+], tester or, as here, the target. Control sequences are referred to as the [−] or, as here, the driver population(s). The flexibility and power of available methods for construction of libraries allow virtually any condition of interest to be investigated. The choice of starting material should be made carefully. It must be remembered that associated phenomena may occur with the condition of interest and for which sequences will also be enriched, for example necrosis occurring with pathogen invasion. The variety of PCR-based techniques available reduce the problems of availability and abundance of starting material.

Subtractive methods have been used to identify differentially expressed genes in plants. These include genes induced by environmental stimuli (Aguan *et al.*, 1991), those involved in response to pathogens (Pautot *et al.*, 1993; Sharma *et al.*, 1993; Wilson *et al.*, 1994) and symbionts (Cook *et al.*, 1995), and genes involved in development (Fujita *et al.*, 1994; Sato *et al.*, 1995). Subtraction methods have been developed for the analysis of genomic DNA (Strauss and Ausabel, 1990), and the technique applied in *Arabidopsis* (Sun *et al.*, 1992).

There are now many accounts for subtractive methods, and their differences indicate that no single method is ideal for all situations. The different methods published aim to avoid or improve one or more perceived difficulty in the technique. In essence there are two strategies, one in which cloned sequences are subtracted, the other in which subtraction occurs before cloning. The decision of which strategy to follow will depend on the availability of material, libraries and personal preference. Both have advantages and disadvantages. The methods are outlined below, followed by some general considerations of the technique.

2.1.1 *Outline of subtractive library construction*

Target and driver sequences are isolated and manipulated in such a way that following hybridization each population can be physically separated from the other. For the hybridization, sequences will usually be single stranded, and driver and target are complementary. The nucleic acids are mixed at high nucleic acid concentration, usually with an excess of driver sequences. Solution hybridization conditions are optimized to allow complementary sequences to anneal. Unhybridized target sequences are isolated and rescued by cloning or transformation.

2.1.1.1 *Clone then subtract*

Early reports describe subtraction being carried out on preconstructed cDNA libraries. This approach should be considered if suitable pre-existing libraries are available or if the library will have other uses. cDNA library construction is described in Chapter 1. Available methods are very efficient and able to capture reliably cDNAs corresponding to rare and long mRNAs. Many phage vectors, for example λ Zap®, allow directional cloning, carry M13 origins of replication and bacteriophage promoters for the facile production of single-stranded DNA *in vivo* or RNA *in vitro*. Although more time-consuming initially it provides a valuable resource for further work. This approach has the advantage of not requiring the cloning of the small quantities of subtracted product, nor the bias created by repeated rounds of PCR. The library will provide essentially unlimited material for the subtractions and, in particular, the large quantities of driver usually required. Library construction can be designed to facilitate the synthesis or production of single-stranded, complementary sequences.

2.1.1.2 *Subtract then clone*

Methods have been described to take advantage of the ability of PCR to amplify cDNA sequences synthesized from very limited amounts of tissue for the construction of cDNA libraries (Jepson *et al.*, 1991), for the production of templates for direct subtraction (Sharma *et al.*, 1993) and for amplification of subtracted DNA prior to subsequent cloning. These approaches are very rapid, and adaptable for numerous subtractive strategies. Possible problems include the potential biased amplification of sequence populations, with the preferential amplification of short sequences and with non-random amplification of sequences (Owens *et al.*, 1991). There are also problems associated with subcloning of PCR-amplified subtracted products (Timblin *et al.*, 1990). There are now commercially available products utilizing a PCR-based approach to subtractive cloning, e.g. PCR-Select cDNA subtraction kit (Clontech).

2.1.2 *Hybridization*

Hybridization conditions are critical to the technique. Accounts that treat the subject of nucleic acid hybridization in detail include Young and Anderson (1985). The overall hybridization strategy is determined by the final desired outcome, which will usually be the elimination from the target of all sequences common to the driver.

Subtractive hybridization typically involves great sequence complexity of the populations and low concentration of many of the constituent sequences. This requires that hybridization is carried out at high $C_o t$ to drive the reaction to completion, where C_o is the original concentration of nucleotides in moles per litre and t is the time in seconds ($C_o t$ of 1 is equivalent to a DNA concentration of 83 µg/ml incubated for 1 h, Young and Anderson, 1985). A $C_o t > 1000$ may be used, a value clearly achieved only by high DNA concentration and extended incubations. Practically, there is a limit to the solubility of DNA and RNA and significant thermal degradation may be expected over extended incubations at temperatures likely to be between 60–75°C. This effect may be reduced by the addition of agents such as formamide which effectively reduce the T_m but at the expense of a reduction in reaction rate. There are a number of methods of increasing the rate of hybridization by increasing the effective DNA concentration. Methods particularly used for subtraction are two-phase systems, especially phenol–water in which nucleic acid adsorption at the emulsion interface markedly increases the reassociation rate (e.g. Sive and St. John, 1988).

The rate of hybridization may be reduced by orders of magnitude if the rate constants for the self-annealing of double-stranded sequence reactions are larger than for hybridization. For this reason it is usual for the populations to be single stranded. Although there are differences in the observable rates of hybridization between RNA:RNA and RNA:DNA, for practical purposes these differences are ignored (Britten and Davidson, 1985).

2.1.3 *Strategies*

The degree of enrichment of target sequences is a function of the driver:target ratio. This has been exploited by Fargnoli *et al.* (1990), who used low ratios at high concentration to drive the reaction to completion in order to enrich for the cDNAs of transcripts that were increased only several fold. More typically, it is desired to enrich for sequences that are greatly induced in the target population compared with the driver population. In such cases multiple rounds of subtraction may be carried out. Milner *et al.* (1995) describe a theoretical model for this situation. Wieland *et al.* (1990) repeated the subtraction of the target fraction three times to obtain up to 700-fold enrichment of desired target sequences. The degree of enrichment or reduction can be monitored by spiking the target with a known sequence (Sun *et al.*, 1992) or by measuring the abundance of genes known to be consecutively expressed in both driver and target (Cecchini *et al.*, 1993).

Theoretically it is possible to hybridize a number of times at high driver:target ratio to leave only sequences unique to the target population. Practically, however, this does not seem to be possible, owing to the much less than ideal actual kinetics and the required $C_o t$ value not being achievable. Reasons include the very varied lengths of sequences present in solution, high concentration of abundant sequences in the driver, the low solubility of DNA, and chemical and thermal degradation of both driver and target sequences. The complexity and distribution of sequences present are unlikely to be known very accurately.

2.1.4 Normalization

In any population of mRNA a few species can be relatively very abundant, for example those for seed storage proteins or rubisco small subunit. Equalization (or normalization) is the process of reducing the fractional representation of abundantly expressed transcripts (Ko, 1990). This has the effect of increasing the fractional representation of less abundant transcripts and has a number of benefits for subtractive hybridization. The increased concentration enables higher $C_o t$ values for less abundant species to be achieved and allows a reduction in the number of subtractive cycles. The technique can increase the detection of sequences that are induced only several-fold irrespective of abundance and improve the detection of rare transcripts by differential screening of cDNA libraries.

2.1.5 Separation of target, driver and hybrids

A critical feature of all protocols is that unhybridized target sequences can be separated from the rest of the hybridization. A variety of methods have been used to accomplish this. One strategy is that only the target sequences are readily clonable. This is accomplished by using intact library clones or their derivatives or by adding specific linkers to the target molecules for subsequent amplification and cloning. Additionally the driver sequences possess tags for their separation, e.g. biotin label, or a polyA tail. They may also be unclonable, typically derivatives that lack specific amplification or cloning sequences, e.g. *in vitro* transcripts or poly[A]$^+$ RNA.

Hydroxyapatite has been used to separate double-stranded from single-stranded sequences (Britten *et al.*, 1974) and silica-based methods have been described (Beld *et al.*, 1996). Most current methods employ biotin labelling of the driver. This allows the use of techniques exploiting the very high affinity of biotin binding to capture hybrids and unhybridized driver, e.g. streptavidin–phenol:chloroform (Sive and St. John, 1988), avidin affinity columns (Duguid *et al.*, 1988) and streptavidin-coated paramagnetic beads. The ease of use and efficiency of these methods has resulted in their widespread use.

Oligo (dT), particularly when attached to a solid support, has been used to simplify and improve subtractive methods. There are now numerous published methods based on this very adaptable technology, with the principal advantage that all manipulations and separations are easily and rapidly carried out directly on the support (Rodriquez and Chader, 1992; Coche *et al.*, 1994, Meszaros and Morton, 1996). This technology can be used to recycle or regenerate sequences for further rounds of subtraction or probe production.

2.1.6 *Cloning of subtracted cDNA*

Clearly if cloned sequences are used in the subtraction, no further cloning is required. The major difficulty may be the very low levels of DNA to be rescued by trans-formation. Very high efficiency transformations can be achieved by electroporation (Dower *et al.* 1988). Alternatively, methods for the preparation of highly competent cells are described in Sambrook *et al.* (1989) and Tang *et al.* (1994). Methods for the concentration of very dilute DNA solutions have been described by Hengen (1996).

The alternative approach requires that the unhybridized target sequences are cloned following subtraction. This invariably requires PCR to generate sufficient DNA for cloning. Linker primers can be designed to minimize non-specific ampli-fication and to allow the use of the most appropriate PCR conditions. Nested amplification can be used to increase the specificity of the final products. Linker ligation is conveniently carried out prior to the subtraction and usually immediately following second-strand synthesis. Biased amplification towards small products can be minimized by size selection of the template prior to PCR and cloning. Size selec-tion based on elution following gel electrophoresis is described in Sambrook *et al.* (1989). Alternative methods are based on size exclusion chromatography, for which there are excellent commercial products, e.g. cDNA spun columns (Pharmacia). The cloning of PCR products can be problematical. However, there are solutions based on the incorporation of endonuclease recognition sites within the PCR primers, the polishing of product ends and the design of cloning vectors. Commercial kits are available for high efficiency PCR product cloning, e.g. TA cloning® system (Invitrogen).

In this chapter, protocols for each of the two different subtractive approaches are described. The first method, based on subtraction of a constructed library, is robust and individual stages are easy to oversee. The second method, based on a subtract then clone strategy, has been used with great success. Details for both are given together with control steps and a troubleshooting section.

2.2 Clone then subtract: subtraction of a phagemid cDNA library

In outline, both target and driver cDNA libraries constructed in λ Zap™ are excised as ssDNA-phagemids. Driver phagemids are converted into biotinylated ssDNA insert sequences by PCR followed by asymmetric PCR (aPCR). Target phagemids are hybridized with this driver DNA at high C_ot. Biotinylated driver and driver:tar-get hybrids are removed with streptavidin paramagnetic beads (stPMB) and the remaining unhybridized target phagemids are rescued by transformation. Transformants are screened for inserts, and insert-containing clones analysed for expression. A flowchart is depicted in Figure 2.1.

2.2.1 Excision of λ cDNA libraries

It is necessary to excise the entire library to ensure complete representation of all sequences. The cell strain must carry F′. The method described below is a modifica-tion of Owens *et al.* (1991).

Target cDNA Library   Driver cDNA Library

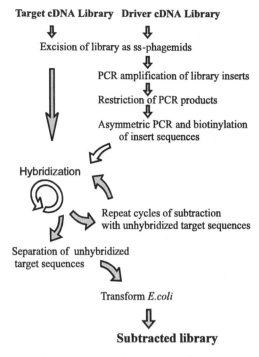

Figure 2.1 Flow diagram for 'clone then subtract' method.

PROTOCOL 2.1   EXCISION OF λ LIBRARIES AS ssDNA PHAGEMIDS

1. Dilute an overnight culture of *E.coli* XL1-blue (Stratagene) 1/100 into 2 ml 2YT with 12.5 µg/ml tetracycline (2TY/T), and grow at 37°C to an $A_{600} = 0.4$ ($\sim 5 \times 10^8$ cells). All subsequent incubations are at this temperature.

2. Co-infect with 20 µl VCS-M13 at 1.0×10^{11} pfu/ml and 20 µl cDNA library at 5×10^9 pfu/ml, and incubate for 10 min.

3. Add 5 ml 2YT/T and incubate with shaking for an hour. All subsequent incubations are with shaking.

4. Add 500 ml prewarmed 2YT/T and incubate for a further hour.

5. Add kanamycin to 25 µg/ml and ampicillin to 100 µg/ml and continue the incubation overnight.

6. Pellet cells at 6000 rpm for 20 min in a Sorval GSA rotor.

7. Precipitate phage from the supernatant with 125 ml 3.5 M ammonium acetate, pH 7.5 with 20% PEG-6000, at 4°C for 2 h and collect by centrifugation at 10 000 rpm at 4°C for 30 min.

8. Resuspend the pellet in 5 ml TE with 200 µg/ml RNase A, and incubate at 37°C for 30 min.

9. Phenol/chloroform extract three times, and chloroform extract once before precipitation with 0.25 M Na acetate/, 2.5 volumes ethanol. Wash the pellet twice with 70% ethanol and resuspend in 250 µl TE.

Phagemid ssDNA quantity and quality can be assessed by gel electrophoresis and by the efficiency of transformation of *E.coli*. The expected yield is around 100 ng ssDNA/ml culture with approximately 20:1 copies of phagemid:helper DNA. Gel electrophoresis shows two major bands, the smaller corresponding to the phagemid and the larger to the helper phage. The smear between these bands is insert-containing phagemid DNA. Electroporation of the ssDNA (Protocol 2.10) should yield $> 10^6$ transformed colonies/µl library.

2.2.2 *Preparation of driver DNA*

The driver DNA is prepared so as to be complementary to the target, single stranded with minimal common vector sequences, and incorporating a tag to allow facile separation from unhybridized target DNA. This is accomplished by symmetric PCR, restriction endonuclease digestion to minimize the inclusion of vector sequences, followed by aPCR incorporating biotin-UTP to generate single-stranded biotinylated driver DNA. The primers are specific to the vector, T7 and T3 for λ Zap. They anneal to the vector close to the cloning site, are highly effective in the amplification and allow nested amplification with other primers in the subsequent aPCR. Alternatively, primers can be used that flank the cloning sites. Annealing temperatures should be calculated or empirically determined for the primer pairs. PCR is carried out for 20 cycles, sufficient for amplification but minimizing the possible preponderant production of vector MCS sequences.

PROTOCOL 2.2 PCR OF PHAGEMID ssDNA

1. Amplify 0.1–1 µl excised driver DNA, in a 50 µl reaction, containing 1 × supplied *Taq* buffer, 0.25 mM dNTPs, 1.5 mM MgCl$_2$, 50 pmol primers and 2 units *Taq* DNA polymerase.

Cycle conditions are:

  94°C × 2 min

  [94°C × 1 min, 50°C × 1 min, 72°C × 2 min] × 20

  72°C × 5 min

2. Following PCR, remove oil and precipitate the products with 0.3 M Na acetate, 2.5 volumes ethanol. Following centrifugation the pellet is washed in 70% ethanol before resuspension in 50 µl TE.

The reaction when analysed by agarose gel electrophoresis, will show a band at ~200 bp corresponding to vector MCS, in front of a smear extending to >1000 bp.

2.2.3 *Restriction endonuclease digestion*

This step is included to reduce the extent of vector sequences carried by the inserts. This step can be omitted if the above PCR was carried out using cloning-specific primers. The chosen enzyme should act close to the original 3' cloning site. The 5' vector MCS sequence is not digested as in the subsequent asymmetric PCR; a specific 5' cloning primer is used resulting in minimal vector sequence amplification.

PROTOCOL 2.3   ENDONUCLEASE DIGESTION OF PCR PRODUCTS

1. An entire purified PCR reaction is digested in a 100 μl volume with 50 units of *Xho*I, in supplied 1 × buffer.

2. Denature enzyme at 65°C for 15 min.

3. Primers, enzymes and small restriction fragments are removed using the Wizard™ DNA CleanUP system (Promega).

2.2.4 *Asymmetric PCR of driver DNA*

Digested, amplified DNA is subjected to aPCR in the presence of a biotin analogue to give a tagged, complementary, essentially single-stranded product. The primer used is 5′ cloning specific, (5′-GGAATTCGGCACGAG). The yield is low as there is no exponential component to the reaction. The size range should extend up from ∼50 bp (ssDNA) but this is not as important as the final representation of sequences. Twenty-five cycles is a reasonable compromise between yield and preponderant synthesis of particular sequences.

PROTOCOL 2.4   ASYMMETRIC PCR OF DRIVER DNA

1. A volume of 1–10 μl template is amplified in a 50 μl reaction containing 1 × supplied *Taq* buffer, 50 pmol primer, 0.25 mM A-,C- and GTP, 0.23 mM TTP, 15 μM biotin-16-dUTP and 2 units *Taq* DNA polymerase.

Cycle conditions are:

    94°C × 2 min

    [94°C × 1 min, 50°C × 1 min, 72°C × 2 min] × 20

    72°C × 5 min

2. Remove unincorporated biotin-nucleotides using the Wizard™ PCRprep system (Promega) following the manufacturer's protocol. The final volume is 50 μl TE.

Incorporation of biotin can be qualitatively determined by use of antibiotin antisera and by binding to stPMB, and the ssDNA content by S1 nuclease digestion.

PROTOCOL 2.5   DETECTION OF ssDNA BY S1 NUCLEASE DIGESTION

1. Digest 5 μl purified aPCR product, in 25 μl S1 nuclease buffer with 1 unit enzyme.

2. Incubate at 37°C for 15 min.

3. Stop with 1 μl 0.5 M EDTA.

Gel analysis should show either an absence of DNA or a gross reduction in its size compared with undigested control, indicating the original presence of ssDNA.

A high DNA yield may indicate that symmetric PCR is occurring, due to the presence of contaminating primers. Re-purify the template to remove these primers. dsDNA can be detected by digestion with frequent cutting restriction enzymes, a reduction in the apparent size of biotinylated products indicating the original presence of dsDNA.

PROTOCOL 2.6 SEROLOGICAL DETECTION OF BIOTIN INCORPORATION

1. Following gel electrophoresis of 1 μl of the DNA, the gel is blotted onto nitrocellulose (Sambrook *et al.* 1989).

2. Block the filter in PBS/5% skimmed milk powder for 60 min, then wash 3 times in 1 × PBS.

3. Incubate the filter in 5 ml PBS containing 1/1000 antibiotin–peroxidase conjugate for 90 min then rinse 3 times in 1 × PBS.

4. Detect the conjugate with 5 ml peroxidase substrate.

5. Stop the reaction with water washes.

The dark blue colour reaction indicating biotin should develop within minutes. Incorporation should be evident as a smear from 200 to 1000 bp.

PROTOCOL 2.7 DETECTION OF BIOTIN INCORPORATION BY STREPTAVIDIN BINDING

Prepare stPMB as described by the manufacturers (e.g. Dynal). This step is helpful in establishing the capacity of the beads for binding ssDNA in later separations. It is absolutely essential to remove all unincorporated biotin-nucleotides following PCR, to prevent the swamping of the streptavidin beads.

1. Add 10 μl 1 M NaCl to 10 μl aPCR reaction.

2. Incubate 10 μl of this solution with 0.1 ml prepared stPMB rotating, for 30 minutes at room temperature.

3. Collect the beads, remove the supernatant to a fresh tube.

4. Analyse by gel electrophoresis (both the supernatant and the untreated control). Any DNA marker used should also be 0.5 M with respect to NaCl.

Gel electrophoresis of the supernatant shows an absence of DNA compared with an untreated control because the biotinylated DNA is bound by the stPMB beads, demonstrating the incorporation of biotin into the driver DNA. The beads do not bind non-biotinylated DNA. Titrate the biotinylated PCR products against the stPMB to determine the capacity of the beads to bind the driver DNA.

2.2.5 *Hybridization*

The concentrations of the driver and target DNAs can be determined by A_{260}. Assume the DNA to be of the average size determined by gel analysis to calculate the concentrations. Adjust the concentrations to reflect the desired ratio of driver: target. In this protocol the ratio of target:driver is 1:20. This can easily be varied, but this is a useful starting ratio.

PROTOCOL 2.8 HYBRIDIZATION

1. A volume of 10 μl [*x* pfu] target and 10 μl [20*x* pfu equivalents] driver DNA are mixed and denatured at 95°C for 5 min.

2. Snap cool on ice and (carefully!) just dry the solution under vacuum.

3. Resuspend the pellet in 3 μl water with gentle vortexing at room temperature for 10 min.

4. Add 1 μl of 4 × hybridization buffer, seal the solution in a glass capillary tube.

5. Hybridize for 24 hours at 65°C.

PROTOCOL 2.9 SEPARATION OF HYBRIDS FROM TARGET DNA

Hybrids and excess driver DNA are removed with an aliquot of stPMB previously shown to be sufficient to remove all the biotinylated driver DNA present in the hybridization.

1. Eject the hybridization solution into an Eppendorf tube containing 100 μl 1 M NaCl and 100 μl of prepared stPMB.

2. Incubate the tube with rotation, for 30 minutes at room temperature.

3. Magnetically separate the beads from the supernatant for 2 min and transfer the supernatant.

4. Wash the stPMB twice with 100 μl 1 M NaCl repeating step 3.

5. Combine the three supernatants and precipitate the target DNA with 2.5 volumes ethanol at −70°C for 1 h, centrifuge for 20 min at 10 000 g, and wash the pellet twice with 70% ethanol before resuspension in 20 μl water. Store at −20°C.

2.2.6 *Electroporation of subtracted library*

Use cells that allow blue-white selection and are F⁻ to reduce helper phage contamination, e.g. *E.coli* DH5. It is essential that the target DNA has a very low salt concentration for electroporation.

PROTOCOL 2.10. PREPARATION OF ELECTROCOMPETENT CELLS

1. Dilute 5 ml of an overnight culture of *E.coli* DH5 into 500 ml SOB and grow with vigorous aeration at 37°C to $A_{600} = 0.3$.

2. Pellet the cells at 3000 rpm for 10 min at 4°C. All subsequent steps are carried out at 4°C.

3. Resuspend the cell pellet in 500 ml 1 mM HEPES pH 7.0, and re-pellet.

4. Sequentially wash the cells (step 3) with 500 ml water, 250 ml water, and then 25 ml 20% glycerol before final resuspension in 1 ml 20% glycerol.

5. Flash-freeze 50 μl aliquots in dry ice/acetone and store at −70°C.

The cell pellet becomes less tightly packed with subsequent washings. Cells should have transformation efficiencies of >10^9 colonies/μg dsDNA.

PROTOCOL 2.11 TRANSFORMATION OF SUBTRACTED LIBRARY

1. Electroporate 1 μl aliquots of library, diluted as necessary, into a 50 μl aliquot of electrocompetent cells. Settings are 1.8 kV, 200 Ohms, 25 μFarad for 0.1 ml cuvettes.

2. Allow the cells to recover at 37°C in 1 ml SOB for 1 h.

3. Pellet the cells, in a microfuge for 10 s, resuspend in 100 µl LB and spread onto LB/ampicillin (100 µg/ml)/Xgal (25 µg/ml)/IPTG (50 µg/ml) plates.

4. Incubate plates at 37°C overnight.

Using the above regime a subtraction efficiency of 80–90% relative to a control of target minus driver has been obtained. This value can be adjusted by varying the proportion of driver to target or by repeated subtraction. Analysis of subtracted clones is described in Section 2.4. and Chapter 7.

2.3 Subtract then clone: PCR-based subtractive cloning

In the following protocol, oligo(dT)$_{25}$-PMB are used for the isolation of mRNA, as a support for subsequent manipulations and for the separation of unhybridized sequences from the hybridization. PMB-bound target and driver mRNA are used as templates for first-strand synthesis. Specific linkers are added to target sequences for subsequent specific PCR amplification and they are then converted into second-strand sequences. Target cDNAs as solution second-strand sequences are hybridized with driver first-strand PMB. Following separation of unhybridized target sequences, PCR is used to amplify larger quantities for subsequent cloning. A flow chart for the method is shown in Figure 2.2.

2.3.1 *Isolation of mRNA*

There are published accounts for the isolation of RNA and mRNA from a variety of plant species and tissues that can be followed. A compilation of methods is included in Croy *et al.* (1993) and commercial systems for the isolation of RNA and mRNA are widely available. The integrity of the RNA/mRNA template is critical for the synthesis of the high quality cDNA on which these protocols depend. It is essential that precautions are taken to remove and exclude nucleases during the procedures. A measure of the purity of the RNA can be gauged from the ratio of A_{260}/A_{280} with values of ≥ 1.7 being acceptable. As a guide, assume the mRNA content to be 1% of total RNA, and 5 µg of poly[A]$^+$ mRNA are required. It is not necessary to isolate mRNA from total RNA as this is effectively accomplished by binding to the oligo(dT) PMB.

PROTOCOL 2.12   PREPARATION OF OLIGO(dT) PMB

1. Resuspend 250 µl oligo(dT) PMB, collect, and remove the supernatant.

2. Wash twice in 200 µl of 2 × binding buffer.

3. Resuspend in 100 µl 2 × binding buffer.

PROTOCOL 2.13   BINDING OF POLY [A]$^+$ mRNA TO OLIGO(dT) PMB

1. Adjust 5 µg mRNA or 2 mg RNA into 100 µl 2 mM EDTA.

2. Denature RNA at 65°C for 2 min and ice quench.

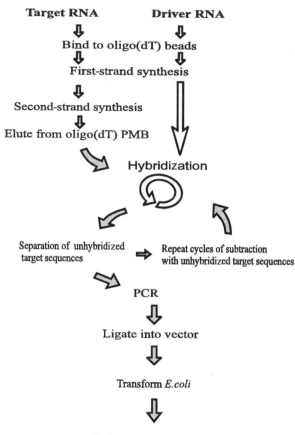

Figure 2.2 Flow diagram for 'subtract then clone' method.

3. Add to prepared oligo(dT) PMB, and anneal at room temperature for 10 min.
4. Wash oligo(dT) PMB twice with 200 µl of 1 × washing buffer
5. Wash in 100 µl × RT buffer.
6. Resuspend in 50 µl 2 × RT buffer.

PROTOCOL 2.14 PHOSPHORYLATION OF THE LINKERS

The linker sequence can incorporate specific sequences to facilitate cloning. Prior to ligation to the target, the linkers require phosphorylation.

1. Add the following reagents:

    5 µg linker at 0.5 µg/ml

    3 µl water

    2 µl 1x supplied polynucleotide kinase (PNK) forward buffer

    3 µl 10 mM ATP

    20 units T4 PNK

2. Incubate at 37°C for 60 min.

3. Pool 10 µl each of oligo 1 and oligo 2, heat to 70°C for 10 min and cool slowly to allow annealing.

4. Store at −20°C.

PROTOCOL 2.15 cDNA FIRST-STRAND SYNTHESIS

Synthesis of cDNA can be carried out directly and conveniently on the mRNA:oligo(dT) PMB. Following cDNA synthesis target and driver are treated differently.

1. Add the following components of the reaction to each of the driver and target mRNA oligo(dT) PMB:

    10 µl 10 mM DTT

    10 µl 30 mM MgCl$_2$

    10 µl 5 mM dNTPs

    10 units RNase inhibitor

    water to 95 µl

    5 µl 200 units/µl Superscript II^{TM}.

2. Incubate, rotating for 1 h at 42°C.

PROTOCOL 2.16 PREPARATION OF DRIVER PMB

Before the driver is used in the hybridization, the RNA must be removed. The eluted mRNA can be later analysed by gel electrophoresis.

1. Wash driver PMB, and resuspend in 100 µl 5 × SSC.

2. Heat the driver PMB to 95°C for 2 min and ice quench.

3. Collect the driver PMB and remove the supernatant.

4. Wash with 200 µl 5 × SSC, remove the supernatant, combining it with the first.

5. Resuspend the driver PMB in 20 µl TE and store at −20°C.

2.3.2 *Preparation of target sequences*

The target is used in the hybridization as second-strand (complementary) products to which have been ligated linkers for subsequent PCR and cloning. Second-strand synthesis is carried out on the target-PMB. Commercial kits can be adapted for use.

PROTOCOL 2.17 TARGET cDNA SECOND-STRAND SYNTHESIS

1. Wash the target-PMB into 138 µl ice-cold water (final volume 200 µl).

2. Add the following reagents:

    40 µl 5 × SS buffer

    2 µl 5 mM dNTPs

    6 µl 5 mM β-NAD+

35

4 units RNase H

20 units *E.coli* DNA ligase

100 units *E.coli* DNA polymerase 1

3. Incubate, rotating at 12°C for 1 h, then at 22°C for 1 h.

PROTOCOL 2.18 COMPLETION AND BLUNT-ENDING OF TARGET cDNA

This step allows completion of second-strand synthesis and the blunt-ending of cDNA to allow efficient ligation of the linkers.

1. Collect the target-PMB, remove the supernatant and resuspend in 40 μl water (final volume 200 μl).

2. Add the following reagents:

   20 μl 5 × SS buffer

   6 μl 5 mM dNTPs

   20 μl 5 mM β-NAD+

   1 μl 10 μg/ml RNase A

   4 units RNase H

   20 units *E.coli* DNA ligase

   10 units T4 DNA polymerase

3. Incubate, rotating at 12°C for 1 h, then at 22°C for 1 h.

4. Collect the target-PMB, remove the supernatant.

5. Wash the target-PMB twice in TE.

6. Resuspend the target-PMB in 50 μl TE and store on ice.

PROTOCOL 2.19 LIGATION OF LINKERS TO TARGET cDNA

1. Collect the target-PMB and wash into 17.5 μl water.

2. Add the following reagents:

   5 μl 10 × supplied T4 ligase buffer

   2.5 μl 10 mM ATP

   5 μl T4 DNA ligase

   20 μl of annealed phosphorylated oligo 1:oligo 2

3. Incubate, rotating, overnight at 16°C.

PROTOCOL 2.20 ELUTION OF COMPLETED TARGET SECOND-STRAND SEQUENCES

The first washes remove all primers and subsequent heating releases the second-strand products into a buffer suitable for hybridization. The target-PMB are kept for further second-strand synthesis.

1. Collect the target-PMB and wash three times with 200 μl TE.

2. Wash once with 5 × SSC and then resuspend in 200 μl 1 × hybridization buffer.

3. Heat target-PMB to 95°C for 2 min and ice quench.

4. Transfer the supernatant to a fresh tube.

5. Repeat the wash with 200 µl 1 × hybridization buffer and pool the supernatants.

6. Resuspend the target-PMB in 50 µl TE and store at −20°C.

A 10 µl sample of the supernatant can be retained for later analysis.

PROTOCOL 2.21 HYBRIDIZATION

Hybridization conditions can be varied. Short incubation times, e.g. 2 h, allow the hybridization of common abundant sequences; longer times also allow the hybridization of rare common sequences. Further rounds of hybridization are accomplished by returning eluted unhybridized target supernatant to the driver-PMB (this protocol, step 8).

1. Collect the driver-PMB and remove the supernatant.

2. Add the eluted target cDNA, ∼390 µl.

3. Heat to 95°C for 5 min, cool slowly to 65°C.

4. Hybridize, rotating for 24 h at 65°C.

5. Collect the beads, transfer the supernatant to a fresh tube.

6. Resuspend the driver-PMB with 200 µl 5 × SSC, heat to 95°C for 2 min, ice quench and remove supernatant to a fresh tube.

7. Repeat the last wash, including the heating step, and pool the two supernatants.

8. Resuspend the driver-PMB in 20 µl TE and store at −20°C.

Prior to PCR the target sequences should be purified by passage through a cDNA spun column or similar, following the manufacturer's protocol.

PROTOCOL 2.22 PCR OF SUBTRACTED TARGET cDNA

PCR is used to generate a larger population for subsequent cloning, and for analysis of target DNA product size distribution and yield. To obtain sufficient material for cloning, the subtracted fraction may be increased by bulk PCR. The optimal template concentration will have to be determined by trial. The primers used are complementary to the ligated linker and the poly(dA) tract.

Amplify 0.1–10 µl target DNA, in a 50 µl reaction, containing 1x supplied *Taq* DNA polymerase buffer, 1.5 mM $MgCl_2$, 0.25 mM dNTPs, 50 pmol primers and 2 units *Taq* DNA polymerase. The primers are oligo 1 and oligo 3.

PCR cycle conditions are:

94°C × 1 min

[94°C × 1 min, 30°C × 1 min, 72°C × 3 min] × 2

[94°C × 1 min, 50°C × 1 min, 72°C × 3 min] × 8

[94°C × 1 min, 50°C × 1 min, 72°C × 4 min] × 10

[94°C × 1 min, 50°C × 1 min, 72°C × 5 min] × 10

72°C × 5 min

The final subtracted product should be size fractionated to reduce the preponderence of small-size PCR products and simultaneously purify the products. This is achieved by cDNA spun column or similar. Alternatively, particular size ranges can be obtained by specific elution following agarose gel electrophoresis (Sambrook *et al.* 1989).

2.3.3 *Cloning of subtracted cDNA*

If suitable restriction enzyme digestion sites have been incorporated into the PCR primers, these should be used for cloning. Alternatively, there are specialized vectors for the direct cloning of PCR products that can be used, e.g. TA cloning® kit (Invitrogen). Prior to ligation, the restricted PCR products should be purified with a commercial purification system, e.g. Promega Wizard PCR Prep or similar, following the manufacturer's protocol. Ligate into appropriately digested, phosphatased vector that allow blue–white selection. The average size of the inserts should be used to calculate the vector:insert ratio. Use a 3:1 and 1:1 vector:insert ratio. Transform prepared suitable competent cells by electroporation (Protocol 2.11).

2.3.4 *Oligonucleotide sequences*

These sequences possessing *Eco*R1 restriction sites have been used to successfully produce a subtracted library:

Oligo 1, 5'-CTCTTGCTTGAATTCGGACTA
Oligo 2, 5'-TAGTCCGAATTCAAGCAAGAGCAC
Oligo 3, 5'-GCATGAATTCGATGC(dT)$_{15}$

2.4 Library screening

This section describes approaches for the screening and provisional analysis of the subtracted library. An account of differential screening is given in Chapter 1 of this volume. Detailed accounts of other methods are well described, e.g. in Sambrook *et al.* (1989). If the subtraction was at or close to completion there are likely to be few clones. Both target and driver cDNA probes can be used to identify abundant constitutive clones and relatively abundant target sequences. Many clones will not be detected by target cDNA probes because of the very low abundance of the transcripts they represent, see cold-plaque screening (Hodge *et al.* 1992). To detect rare sequences in the subtracted library it should be differentially screened with a probe prepared from a subtracted target cDNA.

2.4.1 *Production of a subtracted target probe*

This is conveniently prepared by PCR-based methods. For the first-strand synthesis (Protocol 2.15), 10 μg of mRNA isolated from target and driver tissue (Protocol 2.13) are used as template. Target DNA is further converted into second-strand sequences (Protocol 2.17). There is no necessity to add linkers. Eluted second-strand products are purified by cDNA spun column and used in the hybridization (Protocol 2.21). Following one or more rounds of subtraction, remaining unhybridized DNA is purified by cDNA spun column and used for probe synthesis as described in Chapter 1, Protocol 1.10.

2.4.2 *Primary screening*

Colony hybridization can be used for preliminary screening of large numbers of colonies. A problem with this method can be a high background which obscures the low signal produced by low-abundance sequences. Similarly the use of slot blotting as a technique, although rapid, can also give high backgrounds. Alternatives, although labour intensive, avoid this problem.

One such method is the isolation and appropriate restriction digestion of individual plasmids. The digestion products are separated by gel electrophoresis, the gel blotted and the resulting membrane hybridized with prepared probes. Plasmid isolation methods suited to the handling of large numbers of samples are described in Sambrook *et al.* (1989). Alternatively, PCR can be used to amplify the insert sequences using vector (or cloning-specific) primers, the products separated by gel electrophoresis, blotted, and hybridized with prepared probes. A suitable method for the amplification of insert sequences is described in Gussow and Clackson (1989). Manipulations can be minimized by the use of 96-well plates for both bacterial cell disruption and PCR. There are further advantages with the latter two approaches. DNA sample concentrations can be accurately estimated and equivalent gel loadings achieved. They enable the visualization of insert sequence size, following gel electrophoresis. Duplicate membranes are easily prepared for subsequent differential screening. Relative quantification of expression is possible by the comparison of gel band and autoradiograph intensities. Sequencing is readily carried out on the products for further information.

2.4.3 *Further analysis*

The origin of the clone should be confirmed by either genomic Southern analysis or genomic PCR analysis. Provisionally identified clones should be further analysed to determine whether they are differentially expressed. Available techniques include Northern analysis, RNase protection (Sambrook *et al.* 1989) and quantitative RT–PCR (Grassi *et al.* 1994). Sequencing is an important tool in the analysis of the differentially expressed clones, comparison with sequence databases identifying novel sequences or alternately spliced transcripts. Shotgun sequencing has been used for the rapid analysis of subtracted clones (Perret *et al.* 1994) and is readily carried out on the purified plasmids and PCR products from the preliminary screening. Full or fuller length transcripts can be isolated from the original target cDNA library by PCR. Further analytical methods are described in Chapter 7.

Screening the library will give a variety of clones, from those that are constitutively abundant to those for which no expression can be detected. Those in between will provide many hours of fun and laughter without worrying about the perfection of the subtraction. Be pragmatic in accepting the results of screening of the subtracted library.

2.5 Troubleshooting

This section describes possible problems that may occur during the protocols described above, and possible solutions to them.

2.5.1 *Clone then subtract*

2.5.1.1 *Excision of λ library*

If problems are occurring, separate infections of either only helper phage or only λ library will confirm the infectivity and replication of these viruses, and the mobility of the phage DNA species. Ensure suitable selection for each control. Check helper phage titre as the efficiency of excision is directly related to titre. If necessary prepare fresh, high-titre phage at 10^{11} pfu/ml. There is considerable variation in the efficiency with which different strains of helper phage effect rescue, and different strains can be tried. The different strains carry different genes for selection and also possess differently sized genomes to M13 KO7. This phage requires propagation with kanamycin selection to prevent reversion to wild type. If this occurs helper phage replication will be at the expense of library rescue. If there is no helper phage, check that appropriate selection is being maintained, otherwise obtain fresh stock of helper phage. Yields of dsDNA and ssDNA can be altered by changing growth conditions, e.g. decrease incubation temperature and increase aeration. Low yields may appear as an almost non-existent smear of different-sized DNAs, each at low concentration over an expanse of gel and, may appear faint in relation to helper. It is possible to rescue the DNA running between helper and vector by gel elution, possibly combining a number of trials, to provide sufficient material for subtraction. Digestion of phagemid DNA with *Not*I can be used to reduce the population of dsDNA pBluescript.

2.5.1.2 *PCR*

To increase yield of PCR products, optimize PCR conditions with a number of isolated phagemids of differing size. In the PCR protocol T3 and T7 primers are used. Other primer pairs or combinations can be used, e.g. reverse and −20, SK and KS, that still allow nested amplification with other primers in subsequent aPCR.

2.5.1.3 *Asymmetric PCR*

High DNA yield in this step may indicate that symmetric PCR is occurring, due to the presence of contaminating primers. dsDNA can be detected by digestion with frequent- cutting restriction enzymes. A subsequent reduction in the apparent size of biotinylated products will indicate the original presence of dsDNA. Re-purify the template to remove these primers. Low DNA yields are expected. If exceptionally low, alter PCR conditions or pool the products from a number of reactions.

2.5.1.4. *Low or no biotin incorporation*

It is probable that the PCR is inefficient rather than there is no biotin incorporation. If incorporation is being monitored with stPMB, it is essential to remove unincorporated biotin-nucleotides from the reaction prior to the test. Alternative methods of biotinylation include PCR with a biotinylated primer, although this will lower the power of separation. If the degree of biotin incorporation is low, the biotin ratio can be increased from 4:1 to 2:1.

Ensure that the stPMB have been prepared according to the manufacturer's instructions prior to use. Unincorporated biotinylated nucleotides must be removed following PCR otherwise they easily saturate the binding capacity of the beads. A 1 mg amount of stPMB should bind 20–40 pmol dsDNA of size 0.25–2 kb. Optimal binding of nucleic acid to the stPMB requires at least 1.0 M NaCl. Long DNA fragments require up to 30 min incubation for effective immobilization onto the beads.

2.5.1.5 Hybridization

If DNA is difficult to dissolve following precipitation, do not allow DNA to completely dry before resuspension. The sample can be heated to 60°C to aid solvation.

2.5.1.6 Transformation

No colonies? Ensure that ionic contamination is minimal, to prevent interference with electroporation. Are the cells competent? Check with dsDNA, efficiency should be $> 10^9$ transformants/µg dsDNA.

Few cDNA-containing transformants may be the result of severe contamination with non-recombinant vector. Select colonies within 24 h, to minimize unwanted satellite growth which may confuse subsequent analyses or overgrow the plate. Depending on circumstances, it may be possible to add further selection, e.g. an extra antibiotic. Confirm that the excised target phagemids do not contain high levels of dsDNA. Digest with a frequent-cutting restriction endonuclease and re-purify, prior to transformation. An excess of vector sequences in the driver DNA can hybridize sufficiently to effectively remove all target sequences.

2.5.1.7 Assessment of transformants

Perform control subtractions using only target phagemids to give the baseline population prior to subtraction. The use of driver aPCR products will indicate whether there is a contamination with transformable driver sequences.

If there are no subtracted clones, then confirm the existence of known differentially expressed, common abundant and common rare sequences, in the original library, in the excised phagemid ssDNA and in the aPCR driver. Confirm that the orientation of the driver and target sequences are complementary.

2.5.2 Subtract then clone

2.5.2.1 mRNA yield and quality

Determine mRNA yield by spectrophotometry and quality by Northern blot analysis with known probes following its elution from the oligo(dT) PMB. Re-prepare if necessary, taking great care to prevent nuclease degradation.

2.5.2.2 First- and second-strand synthesis

Synthesis can be monitored by the incorporation of radiolabel ($[\alpha\text{-}^{32}\text{P}]$ dCTP) followed by gel analysis and autoradiography. Heat a small aliquot of the oligo(dT) PMB to 95°C for 2 min, then ice quench. Remove the mRNA supernatant and use for a parallel first-strand synthesis including $[\alpha\text{-}^{32}\text{P}]$ dCTP. Second-strand synthesis can be monitored in a small representative aliquot of the reaction by the incorporation of radiolabel ($[\alpha\text{-}^{32}\text{P}]$ dCTP).

2.5.2.3 Linker phosphorylation and ligation

Phosphorylation of the linkers can be monitored by tracing the incorporation of $[\gamma\text{-}^{32}\text{P}]$ ATP. This label will subsequently be incorporated into cDNA following linker ligation, eluted from the beads with the second-strand sequences, and low levels will remain following hybridization, providing a good estimate for these reactions.

2.5.2.4 PCR

If problems are suspected test and replace PCR components. From very low starting concentrations of DNA, additional rounds of amplification may be necessary.

2.5.2.5 Cloning of subtracted PCR products

Possible reasons for few transformed colonies include incomplete digestion of PCR products and/or vector. Confirm that the restriction enzymes are active and that they have been inactivated or removed prior to ligation of inserts into the vector. Test and replace the enzymes and buffers for ligation. Ensure the cells are competent and replace if necessary. Try direct cloning of PCR products, e.g. TA cloning system (Invitrogen).

2.5.2.6 No subtracted clones

Confirm the existence of known differentially expressed, common abundant and common rare sequences, in the mRNA and cDNA of both target and driver.

2.6 Solutions

Binding buffer, 2 ×: 20 mM Tris-Cl pH 7.4, 1.0 M LiCl, 2 mM EDTA.
CIAP buffer: 50 mM Tris-Cl pH 9.0, 1 mM $MgCl_2$, 0.1 mM $ZnCl_2$, 1 mM spermidine.
Hybridization buffer 4 ×: 4 M NaCl, 160 mM EPPS (*N*-[2-hydroxyethyl-piperazine],*N*-[3-propanesulphonic acid]) pH 8.25, 20 mM EDTA.
Peroxidase substrate: 4 ml H_2O, 1 ml chloronapthol (5 mg/l methanol), 20 µl hydrogen peroxidase (6% solution in water).
RT buffer (Superscript II^{TM} BRL): 100 mM Tris-Cl pH 8.3, 150 mM KCl.
S1 nuclease buffer: 30 mM Na acetate pH 4.6, 50 mM NaCl, 1 mM ZnCl.

SS buffer 5 ×: 100 mM Tris-Cl pH 7.5, 500 mM KCl, 50 mM DTT, 25 mM MgCl$_2$, 50 mM (NH$_4$)$_2$SO$_4$, 250 µg/ml BSA.
SSC, 20 ×: 3M NaCl, 0.3M tri-sodium citrate, pH 7.0 with HCl.
STE: 0.1 M NaCl, 20 mM Tris-Cl pH 7.5, 10 mM EDTA.
Washing buffer: 10 mM Tris-Cl pH 7.4, 150 mM LiCl, 1 mM EDTA.
Media for the growth of E.coli are described in Sambrook *et al.* (1989).

References

AGUAN, K., SUGAWARA, K., SUZUKI, N. and KUSANO, T., 1991, Isolation of genes for low temperature-induced proteins in rice by a simple subtractive method, *Plant and Cell Physiology*, **32**, 1285–1289.

BELD, M., SOL, C., GOUDSMIT, J. and BOOM, R., 1996, Fractionation of nucleic acids into single-stranded and double-stranded forms, *Nucleic Acids Research*, **24**, 2618–2619.

BRITTEN, R.J. and DAVIDSON, E. H., 1985, Hybridization strategy, in Hames, B.D. and Higgins, S.J. (Eds) *Nucleic Acid Hybridization: A Practical Approach*, pp. 3–15, Oxford: IRL Press.

BRITTEN, R.J., GRAHAM, D.E. and NEUFELD, B.R., 1974, Analysis of repeating DNA sequences by reassociation, *Methods in Enzymology*, **29**, 363–441.

CECCHINI, E., DOMINY, P.J., GERI, C., KAISER, K., SENTRY, J. and MILNER, J.J., 1993, Identification of genes up-regulated in dedifferentiating *Nicotiana glauca* pith tissue, using an improved method for constructing a subtractive cDNA library, *Nucleic Acids Research*, **21**, 5742–5747.

COCHE, T., DEWEZ, M. and BECKERS, M.-C., 1994, Generation of an unlimited supply of a subtracted probe using magnetic beads and PCR, *Nucleic Acids Research*, **22**, 1322–1323.

COOK, D., DREYER, D., BONNET, D., HOWELL, M., NONY, E. and VANDENBOSCH, K., 1995, Transient induction of a peroxidase gene in *Medicago trunculata* precedes infection by *Rhizobium melliloti*, *Plant Cell*, **7**, 43–55.

CROY, E.J., IKEMURA, T., SHIRSAT, A. and CROY, R.R.D., 1993, Plant nucleic acids, in Croy, R. R.D. (Ed.) *Plant Molecular Biology Labfax*, pp. 21–48. Oxford: Bios Scientific Publishers.

DOWER, W.J., MILLER, J.F. and RAGDALE, C.W., 1988, High efficiency transformation of *E.coli* by high voltage electroporation, *Nucleic Acids Research*, **16**, 6127–6145.

DUGUID, J. R., ROHWER, R. G. and SEED, B., 1988, Isolation of cDNAs of scrapie-modulated RNAs by subtractive hybridization of a cDNA library, *Proceedings of the National Academy of Sciences (USA)*, **85**, 5738–5742.

FARGNOLI, J., HOLBROOK. N. J. and FORNACE, A. J., 1990, Low ratio hybridization subtraction, *Analytical Biochemistry*, **187**, 364–373.

FUJITA, T., KOUCHI, H., ICHIKAWA, T. and SYONO, K., 1994, Cloning of cDNAs for genes that are specifically or preferentially expressed during the development of tobacco genetic tumours, *The Plant Journal*, **5**, 645–654.

GRASSI, G., ZENTILIN, L., TAFURO, S., DIVIACCO, S., VENTURA, A., FALASCHI, A. and GIACCA, M., 1994, A rapid procedure for the quantification of low abundance RNAs by competitive reverse transcription-PCR, *Nucleic Acids Research*, **22**, 4547–4549.

GUSSOW, D. and CLACKSON, T., 1989, Direct clone characterisation from plaques and colonies by PCR, *Nucleic Acids Research*, **17**, 4000.

HENGEN, P.N., 1996, Carriers for precipitating nucleic acids, *Trends in Biochemical Sciences*, **21**, 224–225.

HODGE, R., PAUL, W., DRAPER, J. and SCOTT, R., 1992, Cold plaque screening: a simple technique for the isolation of low abundance, differentially expressed transcripts from conventional cDNA libraries, *The Plant Journal*, **2**, 257–260.

Jepson, I., Bray, J., Jenkins, G., Schuch, W. and Edwards, K., 1991, A rapid procedure for the construction of PCR cDNA libraries from small amounts of plant tissue, *Plant Molecular Biology Reporter*, **9**, 131–138.

Ko, M.S.H., 1990, An equalised cDNA library by the reassociation of short double-stranded cDNAs, *Nucleic Acids Research*, **18**, 5705–5711.

Meszaros, M. and Morton, D.B., 1996, Subtractive hybridization strategy using Para-magnetic oligo(dT) beads and PCR, *BioTechniques*, **20**, 413–417.

Milner, J.J., Cecchini, E. and Dominy, P.J., 1995, A kinetic model for subtractive hybridisation, *Nucleic Acids Research*, **23**, 176–187.

Owens, G.P., Hahn, W.E. and Cohen, J.J., 1991, Identification of mRNAs associated with programmed cell death in immature thymocytes, *Molecular and Cellular Biology*, **11**, 4177–4188.

Pautot, V., Holzer, F.M., Reisch, B. and Walling, L.L., 1993, Leucine aminopeptidase: an inducible component of the defence response in tomato, *Proceedings of the National Academy of Sciences (USA)*, **90**, 9906–9910.

Perret, X., Felley, R., Bjourson, A.J., Cooper, J.E., Brenner, S. and Broughton, W.J., 1994, Subtractive hybridisation and shot-gun sequencing: a new approach to identify symbiotic loci, *Nucleic Acids Research*, **22**, 1335–1341.

Rodriquez, I.R. and Chader, G.J., 1992, A novel method for the isolation of tissue-specific genes, *Nucleic Acids Research*, **20**, 3528.

Sambrook, J., Fritsch, E.F. and Maniatis, T., 1989, *Molecular Cloning: A Laboratory Manual*, USA, Cold Spring Harbor Laboratory Press.

Sato, S., Toya, T., Kawara, R., Whittier, R.F., Fukuda, H. and Komamine, A., 1995, Isolation of a carrot gene expressed during early-stage somatic embryogenesis, *Plant Molecular Biology*, **28**, 39–46.

Sharma, P., Lonneborg, A. and Stougaard, P., 1993, PCR-based construction of subtractive cDNA libraries using magnetic beads, *BioTechniques*, **15**, 610–611.

Sive, H. L. and St. John, T., 1988, A simple subtractive hybridization technique employing photoactivable biotin and phenol extraction, *Nucleic Acids Research*, **16**, 10937.

Strauss, D. and Ausabel, F. M., 1990, Genomic subtraction for cloning DNA corresponding to deletion mutations, *Proceedings of the National Academy of Sciences(USA)*, **87**, 1889–1893.

Sun, T.P., Goodman, H. M. and Ausabel, F.M., 1992, Cloning the *Arabidopsis* GA1 locus by genomic subtraction, *The Plant Cell*, **4**, 119–128.

Tang, K., Nakata, Y., Li, H.-L., Zhang, M., Gao, T., Fujita, A., Sakatsume, O., Ohta, T. and Yokoyama, K., 1994, The optimisation of preparations of competent cells for the transformation of *E.coli*, *Nucleic Acids Research*, **22**, 2857–2858.

Timblin, C., Battey, J. and Kuehl, W.M., 1990, Application of PCR technology to subtractive cDNA cloning: identification of genes specifically expressed in murine plasmacytoma cells, *Nucleic Acids Research*, **18**, 1587–1593.

Wieland, I., Bohm, M. and Bogatz, S., 1990, Isolation of DNA sequences deleted in lung cancer by genomic difference cloning, *Proceedings of the National Academy of Sciences(USA)*, **89**, 9705–9709.

Wilson, M.A., Bird, D. McK. and van der Knaap, E., 1994, A comprehensive subtractive cDNA cloning approach to identify nematode-induced transcripts in tomato, *Phytopathology*, **84**, 299–303.

Young, B.D. and Anderson, M.L.M., 1985, Quantitative analysis of solution hybridization, in Hames, B.D. and Higgins, S.J. (Eds) *Nucleic Acid Hybridization: A Practical Approach*, pp. 47–111, Oxford: IRL Press.

Differential Display of mRNA

GREGORY R. HECK AND DONNA E. FERNANDEZ

3.1 Introduction

Differential display of mRNA, first described by Liang and Pardee (1992), is a technique for making broad-scale surveys of transcribed gene expression patterns and subsequently cloning sequences with desired expression characteristics. The technique relies upon the use of arbitrary primers and the polymerase chain reaction (PCR), and thus is similar in concept to previously established techniques such as randomly amplified polymorphic DNA (RAPD) analysis of genomic DNA. Informative patterns or 'fingerprints' of amplification products can be produced even when no previous information is available concerning primer binding sites and/or expected products. These patterns provide the basis for selecting and ultimately isolating differentially expressed genes and have even been suggested as a means for identifying and classifying different RNA sources (Liang *et al.*, 1993).

Using differential display, the large number of genes represented in any given sample of RNA can be systematically examined. To accomplish this, a specific subset of mRNAs is converted to cDNA by reverse transcription with an oligodeoxythymidine primer oligo(dT) anchored at the $3'$ terminus by one or two specified bases (e.g. $T_{12}AC$). A portion of the cDNA is then subjected to PCR amplification using the oligo(dT) from the first-strand synthesis reaction and a second primer of arbitrarily chosen sequence, typically a 10 base oligonucleotide. Annealing and amplification is performed under conditions that favour degenerate priming, at least in the initial cycles of amplification. This typically produces a set of 50–100 amplification products, up to several hundred base pairs in size, which represent a manageable subfraction of the thousands of transcribed sequences in the total mRNA population. Incorporation of a labelled nucleotide during the PCR amplification allows the products to be visualized or 'displayed' by autoradiography after separation on a polyacrylamide sequencing gel. If the process is repeated with one or more additional mRNA populations, the amplification products can be directly compared by electrophoresis in adjacent lanes. Upon inspection, a small number of products may be identified that show induction or repression in the compared mRNA populations. Differentially expressed amplification products can then be excised from the dried gel, eluted, and re-amplified to yield sufficient material for cloning, sequencing, and probe production (see schematic overview Figure 3.1).

3.1.1 Strengths and limitations of differential display of mRNA

Technical simplicity and a requirement for minimal amounts of tissue are significant strengths of the differential display method. All aspects of the protocol can be performed in any laboratory generally equipped for PCR, DNA sequencing and recombinant DNA cloning. Small amounts of total RNA, of the order of a few micrograms from each sample being compared, are sufficient for display of a large number of primer combinations. The low demand for RNA template in differential display opens up the possibility of using tissue sources that cannot be acquired in large quantity.

Sensitivity is another potential advantage of differential display as a tool for cloning differentially expressed genes. Liang and Pardee (1992) demonstrated that the thymidine kinase (TK) transcript, which is estimated to be present at less than 50 copies per cell, can be detected in a display autoradiogram. However, their 'arbitrary' decamer was perfectly complementary to a sequence within the TK transcript, while most amplifications arise from initially degenerate binding of the arbitrary primer. Experiments in *Brassica napus* utilized differential display to isolate a transcription regulator, AGL15 (see Figure 3.2, panel C and Figure 3.3), estimated to constitute 0.02% of the transition-stage embryo mRNA population (Heck *et al.*, 1995). In this instance, initial priming did occur in a degenerate manner with the decamer acting as a hexamer or heptamer, indicating that lower abundance transcripts can be visualized by this technique even without perfect complementarity of

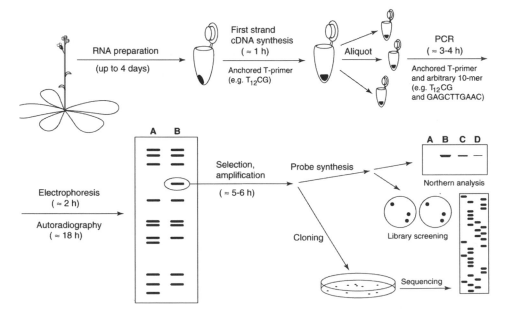

Figure 3.1 Schematic overview of differential display of RNA.

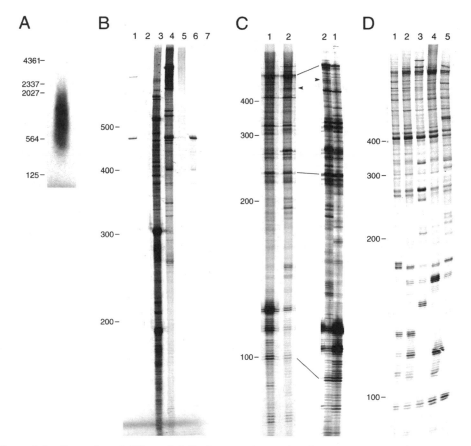

Figure 3.2 Examples of differential display reactions using *Brassica napus* total RNA.

A. Autoradiogram of cDNA synthesized using a $T_{12}GC$ primer and 1 μg of total RNA isolated from *B. napus* embryos, 45 days after pollination (dap). Labelled cDNA was separated on a 1.4% alkaline agarose gel. Size markers, in bases, are indicated to the left of this and all subsequent panels.

B. Autoradiogram of sample reactions and controls using total RNA from *B. napus* embryos 45 dap. (1) Control reaction, PCR amplification with primers $T_{11}GT$ and CGCAGTACTC, using total RNA template (without synthesis of first-strand cDNA by reverse transcriptase). (2) Same control reaction as lane 1 with RNase A added prior to amplification. (3) cDNA synthesis with $T_{11}GT$ followed by amplification with $T_{11}GT$ and GAGCTTGAAC. (4) cDNA synthesis with $T_{11}GT$ followed by amplification with $T_{11}GT$ and CGCAGTACTC. (5) Control reaction, cDNA synthesis with $T_{11}GT$ followed by amplification with $T_{11}GT$ primer alone. (6) Control reaction, cDNA synthesis with $T_{11}GT$ followed by amplification with arbitrary decamer, CGCAGTACTC alone. (7) Control reaction, no template (H_2O only) and amplification with $T_{11}GT$ and CGCAGTACTC.

C. Reproducibility of differential display. Differential display was performed independently by two researchers using the primers $T_{12}CG$ and GAGCTTGAAC and total RNA isolated from vegetative shoot apices (lane 1) and embryos 45 dap (lane 2). Guide lines are given to assist in orientation of reactions between autoradiograms. Arrows indicate the position of the AGL15-1 (also shown in Figure 3.3).

D. Autoradiogram showing the variation in differential display patterns obtained from selected developmental stages and organs using a single primer set. cDNA was synthesized using a $T_{12}GG$ primer on total RNA isolated from embryos at globular/heart transition stage (∼16 dap, lane 1), heart stage (∼20 dap, lane 2), maturation stage (45 dap, lane 3), vegetative shoot apices (lane 4) and mature leaves (lane 5). Amplification was performed with the primers $T_{12}GG$ and GAGCTTGAAC.

primers. Low abundance transcripts, such as AGL15, may not be as readily isolated using more traditional methods such as differential hybridization.

Perhaps the greatest advantage of differential display is that the technique allows an investigator to compare gene expression in multiple samples. Careful selection of RNA sources can help to identify genes that are involved, for instance, in specific aspects of a physiological process or a developmental progression and are not linked to general aspects of plant growth and development. For example, to identify genes involved in early embryo development in *Brassica napus*, five RNA populations were compared. Two were isolated from transition and heart stages, when the young embryo was undergoing morphogenesis (cell division and differentiation), and one was isolated from a late embryo stage, when seed desiccation had already started. To eliminate products related in a general way to the processes of cell division and expansion, a young vegetative shoot apex sample was included. Finally, a sample from mature leaves with fully differentiated cells was displayed to help screen out general housekeeping genes. One primer combination used in this analysis is shown in Figure 3.2, panel D. As is evident in the display autoradiogram, most of the visualized products are found in all samples and only a small number potentially represent genes specific to early embryo development. In this comparison, the most distinctive display pattern is derived from the late stage embryo, which is also the most distinctive sample in terms of physiology and development. Similar selection strategies have been used in other plant systems to obtain genes associated with fruit ripening in strawberry (Wilkinson *et al.*, 1995), response to gibberellic acid treatment in deep-water rice (van der Knaap and Kende, 1995), and dormancy in wild oat (Johnson *et al.*, 1995).

Regardless of the criteria used to select a given sequence using differential display, any experimental strategy must ultimately include a means of independently confirming the expression pattern. Essentially this requires that Northern analysis, quantitative reverse transcription PCR, *in situ* hybridization or other techniques be performed to verify the differential display results. In most cases, the patterns observed in differential display are highly reproducible (an example is shown in Figure 3.2, panel C). However, small quantities of starting template, relaxed priming conditions, and the exponential nature of PCR can all contribute to bias in the representation of individual sequences, making differential display less reliable, in a quantitative sense, than Northern analysis. For example, the T6 sequence was initially chosen from a display gel of *Brassica napus* mRNAs because it appeared to be restricted to early embryogenesis (Figure 3.3, panel A). When the re-amplified and radiolabelled T6 PCR product was used as a probe on a Northern blot (Figure 3.3, panel B), expression was noted in the young embryo undergoing morphogenesis but also in somewhat older embryos and, unexpectedly, in the vegetative shoot apex. Factors extrinsic to the technique, such as expression of closely related gene family members or selection of contaminating comigrating sequences from the display gel, could also account for this and other unexpected patterns, (see section 3.3).

Before undertaking a large scale analysis like differential display, an investigator should seriously consider if this technique is most appropriate for his or her application. Differential display of mRNA is inappropriate, for instance, for isolating a particular target gene, such as one responsible for a specific phenotype. To display all of the 10 000 or more genes expressed in a given cell, as few as 80 or more than 240 pairwise combinations of oligo(dT) primers and arbitrary decamers may be needed (Bauer *et al.*, 1993; Liang *et al.*, 1993). Although this is possible, it may not be particularly time effective or practical and the ability to sample an mRNA

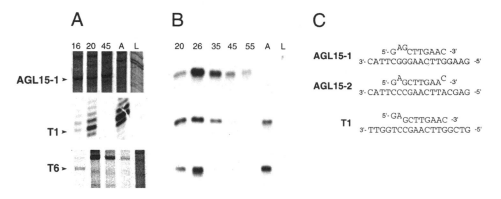

Figure 3.3 Examples of individual differential display products and corresponding expression patterns determined by Northern analysis.
A. Differential display reactions. AGL15-1, T1, and T6 products are indicated by arrows in portions of autoradiograms from amplifications using primer sets $T_{12}CG$ and GAGCTTGAAC, $T_{12}GG$ and GAGCTTGAAC, and $T_{11}AG$ and ATCTCGCTAG, respectively, and *B. napus* total RNA. Lane designations for this panel and panel B: 16–55, embryo age in dap; A, vegetative shoot apices from 2-week-old seedlings; L, mature leaf.
B. Corresponding Northern analyses for the differential display products shown in panel A. Five µg total RNA were used per lane and probes were prepared either directly from re-amplified differential display products (T6) or from clones subsequently isolated from a *B. napus* embryo cDNA library (AGL15-1 and T1).
C. Proposed priming sites for the arbitrary decamers used in the differential display of AGL15-1 and T1, 485 and 166 bp upstream of the polyadenylation sites, respectively. Superscripts indicate mismatches between the primer and cDNA sequence, determined by comparison with sequenced cDNA clones from a *B. napus* embryo cDNA library. The embryo-expressed AGL15-2 (an AGL15-1 homologue) sequence is not primed by GAGCTTGAAC because of a nucleotide mismatch at the 3′ terminus of the arbitrary primer.

population exhaustively has not been demonstrated. The differential display technique is better suited to obtaining genes that are representative of the group affected by a mutation or that take part in a physiological process or developmental progression. The number of differential products that can be expected in any particular screen will vary greatly, depending on the overall similarity of the compared RNA populations. In screens for differentially expressed genes in cancerous cell lines (Liang *et al.*, 1993) or genes involved in regeneration of the mouse liver (Bauer *et al.*, 1993), less than 1% of the bands on the display autoradiograms appeared to be differential. Of these, perhaps 30% could be confirmed by Northern analysis using total RNA. Because the level of false positives (unconfirmed differential expression) is quite high, careful attention must be given to execution of the technique, controls and subsequent analysis of differential display products.

3.2 Methods

3.2.1 Plant resources

Plant material for RNA extraction should be harvested, immediately frozen in liquid nitrogen, and stored at $-70°C$ until reagents and equipment for processing have been assembled. Although other sources are suitable, axenically grown plants of genetically uniform genotype are perhaps the most ideal source of material. Because of the potential of PCR to amplify rare nucleotide sequences, even small amounts of co-isolated RNAs derived from parasites, symbionts or other microflora associated with field grown material may contribute significantly to the final display patterns. Additionally, since the initial amplification is sensitive to even single nucleotide polymorphisms, display patterns may be skewed when different alleles of an expressed gene are present in compared samples (for an analogous example, see the priming mismatches of the AGL15 orthologues in Figure 3.3, panel C). Thus, the investigator should be aware of sources of variation other than differential gene expression in the collected material.

3.2.2 *Preparation of total RNA*

A large variety of RNA preparation protocols exist for plant material. Some are faster or more convenient to use than others (e.g. commercial kits) and some are designed to accommodate tissue sources that present particular difficulties, such as high levels of polysaccharides or phenolics. In the final assessment, any protocol that produces high molecular weight, intact RNA with a spectrophotometric absorbance ratio (A_{260}/A_{280}) between 1.9 and 2.0 is suitable for differential display. Trace quantities of co-isolated genomic DNA undetectable by gel analysis or spectrophotometry represent a potential source of false positives during differential display. The degree of genomic DNA contamination can be assessed by PCR of an RNA sample without prior reverse transcription. A low level of *Taq* polymerase-based reverse transcription (see Figure 3.2, panel B, lanes 1 and 2), which can be eliminated by treatment with RNase, should not be confused with amplification from contaminating genomic DNA. DNase treatment of isolated RNA (step 12, Protocol 3.1) should be considered for any method chosen and may allow rapid or miniprotocols that yield lower purity RNA to be used. As with any procedure utilizing RNA, precautions should be taken to prevent degradation of RNA during preparation and its subsequent storage and use, detailed in Chapter 1, Section 1.2.

The following protocol represents one possible extraction procedure to isolate total RNA for use in differential display. This protocol has been used to obtain RNA from a diverse array of organs and developmental stages in *Brassica napus* (Finkelstein *et al.*, 1985) and from other plant species, including *Zea mays*, *Pueraria lobata* (kudzu) and *Welwitschia mirabilis* (G. Heck, unpublished information). Although this protocol is lengthy, it yields high molecular weight RNA without significant contamination by genomic DNA, thus obviating the need for routine DNase treatment prior to use in the differential display technique. However, some control reactions to test for genomic contamination should always be performed.

PROTOCOL 3.1 PREPARATION OF TOTAL RNA

Prepare in advance the homogenization buffer and buffer-saturated phenol.

The following protocol is appropriate for 1 g of tissue. Larger or smaller quantities of material may be processed by scaling volumes up or down and changing homogenization probes to suit the liquid volume. Tissue quantities as low as 20 mg fresh weight have been processed with this procedure.

1. Add 10 ml of homogenization buffer, 5 ml of buffer-saturated phenol, and 5 µl Antifoam A (a mixture of non-ionic emulsifiers to prevent foaming) to a 50 ml conical centrifuge tube. Add frozen plant material and homogenize immediately with a Polytron® (Brinkmann Instruments, Inc., Westbury, NY, USA) homogenizing probe set at a speed which does not splatter contents out of the tube. Bulky samples should be powdered in a mortar and pestle (pre-chilled with liquid nitrogen) prior to homogenization. Continue homogenization for 1–2 min until the tissue appears well disrupted.

2. Transfer to a capped tube or flask and shake vigorously on a wrist-action or rotary shaker for 15 min.

3. Add 5 ml of chloroform:isoamyl alcohol (24:1). Shake for 10 min.

4. Transfer to a capped glass Corex® tube and centrifuge for 10 min at 12 000 *g* to separate the phases.

5. Remove the aqueous phase, avoiding precipitate at the interface. Transfer it to a clean tube and shake with an additional 10 ml of equilibrated phenol:chloroform:isoamyl alcohol (25:24:1) for 10 min. Centrifuge for 10 min at 12 000 *g*.

6. Remove the aqueous phase, transfer to a clean Corex® tube and add 0.4 ml of 5 M potassium acetate pH 6.0 (0.2 M final concentration) plus 10 ml (2 volumes) of 100% ethanol. Allow to precipitate for several hours to overnight at −20°C.

7. Collect precipitate by centrifugation at 12 000 *g* for 10 min. Rinse the pellet in cold 70% ethanol and dry in a vacuum chamber. Resuspend in 0.5–1 ml of TE. Add potassium acetate to 0.2 M and 2 volumes of 100% ethanol and allow to precipitate for 4 h to overnight at −20°C.

8. Centrifuge, rinse and dry the pellet as before. Resuspend in 0.5–1 ml of TE and add an equal volume of 4 M LiCl. Let precipitate overnight on ice.

9. Centrifuge at 12 000 *g* for 30 min and carefully remove the supernatant as completely as possible (pellet should adhere loosely to the tube wall). Resuspend pellet in 0.5–1 ml of TE and precipitate with 1.5 volumes of 5 M potassium acetate for 4–5 h at −20°C.

10. Centrifuge at 12 000 *g* for 15 min and remove supernatant. Dissolve pellet in 0.5 ml TE and add 1 ml of 100% ethanol. Allow to precipitate for 2 h to overnight at −20°C. Centrifuge at 12 000 *g* for 10 min, rinse the pellet with 70% ethanol and dissolve in 0.5–1 ml of H_2O.

11. Measure RNA concentration by diluting an aliquot and measuring absorbance at 260 and 280 nm. One A_{260} unit = 40 µg/ml of RNA and the (A_{260}/A_{280}) ratio should be close to 2. Check the integrity by heating 2 µg of RNA at 65°C for 10 min, cool rapidly on ice and add 0.1 volume of loading dye. Load onto a 1% agarose gel prepared in 1 × TAE with 0.5 µg/ml of ethidium bromide. The 25S and 18S rRNAs should appear as distinct, tight bands against a much fainter background smear. Samples derived from photosynthetic tissue will contain additional bands due to the chloroplast 23S and 16S rRNAs.

12. If necessary, trace quantities of genomic DNA can be removed by treatment of the total RNA with RNase-free DNase I. Add 1 µl DNase I (1 unit/µl) to 25–50 µg of RNA in 50 µl of 10 mM Tris-Cl pH 8.3, 50 mM KCl, 1.5 mM $MgCl_2$, 1 mM dithiothreitol (DTT), and 0.4 units/µl human placental RNase inhibitor. Incubate for 30 min at 37°C. Extract twice with an equal volume of phenol:chloroform (1:1) and precipitate the aqueous phase with 0.1 volume of 3 M sodium acetate (pH 5.2) and 2.5 volume of 100% ethanol. After 30 min at

−70°C, pellet the RNA in a microfuge for 10 min (16 000 *g*). Rinse pellet with 70% ethanol and resuspend at a concentration of 1 μg/μl or higher in H_2O. Store RNA at −70°C, aliquoting into smaller volumes if frequent use is anticipated.

3.2.3 RNA template for first-strand cDNA synthesis

Total RNA is acceptable for use in the reverse transcription reaction. The oligo(dT) primer effectively selects and primes mRNA (∼1–2% by mass) despite the large excess of rRNA and tRNA in a total RNA population. Polyadenylated (poly [A]$^+$) RNA is a suitable template but does not provide a significant qualitative advantage over total RNA. Further extension of limited samples may be achieved by reducing the amount of total RNA template in the reverse transcription reaction. Only slight variations in display patterns are observed when the amount of input RNA template is varied over an order of magnitude from 0.2–2 μg (Liang *et al.* 1993).

3.2.4 Oligo(dT) primer

The original description of the differential display technique used oligo(dT) primers anchored at the 3′ terminus with two additional bases (e.g. $T_{12}AG$). Addition of these defined bases leads to specificity in priming and amplification of a particular subset of cDNAs. To sample all possible mRNAs, 12 different oligo(dT) primers with 2 specified bases and 12 cDNA synthesis reactions are needed. The number of cDNA synthesis reactions can be reduced to 4 if oligo(dT) primers degenerate at the penultimate base (e.g. $T_{12}VG$, where V = A, C or G) are used. In fact, degenerate priming often occurs even with two specified 3′ bases (Liang *et al.*, 1993). Taking this a step further, Liang *et al.* (1994) demonstrated that a single specified 3′-terminal base works well in the context of a longer primer (e.g. $AAGCT_{11}G$, incorporating a 5′-terminal *Hin*d III restriction site). Thus, the spectrum of suitable oligo(dT) primers covers a broad range. A number of commercial firms are currently offering kits containing multiple oligo(dT) primers that can reduce the expense involved in synthesis of multiple oligonucleotides (e.g. Clontech, Palo Alto, CA, USA; GenHunter, Brookline, MA, USA; and Operon Technologies, Alameda, CA, USA).

PROTOCOL 3.2 PREPARATION OF FIRST-STRAND cDNA TEMPLATE

1. Mix 1 μg of total RNA or 0.1 μg of poly[A]$^+$ RNA - (free of genomic DNA)

    2 μl of 25 μM oligo(dT) primer

    H_2O to 9.5 μl final volume

    Heat at 70°C for 10 min, then cool rapidly on ice.

2. Add 10.5 μl of reverse transcriptase solution. Each 10.5 μl of reverse transcriptase solution contains:

    4 μl 5× reverse transcriptase buffer

    2 μl 0.1 mM DTT

2 μl 200 μM (each) dNTP mix

0.5 μl human placental RNase inhibitor (20 units/μl)

1 μl MMLV reverse transcriptase (100 units/μl)

Add the enzyme to the reverse transcriptase solution shortly before dispensing into tubes containing RNA.

3. Incubate at 37°C for 1 h and then inactivate reverse transcriptase by heating for 5 min at 95°C.

4. Use first-strand template immediately or freeze at −20°C (stable for at least 1 year).

3.2.5 *Arbitrary primer choice*

By varying the arbitrary primer, different subsets of cDNAs can be amplified from a single cDNA reaction (see Figure 3.2, panel B, lanes 3 and 4). Few restrictions are placed on the design of the arbitrary primer, however, uninterrupted stretches of self-complementarity (particularly with the 3′ end of the primer) and repetitive sequences (e.g. GAGAGAGAGA) should be avoided. The arbitrary primer should also have a base composition of approximately 50–60% G + C content. In a decamer, the 6–7 bases at the 3′ terminus are the most critical in the initial rounds of amplification. Hence, when designing a set of arbitrary primers, it is critical that these terminal nucleotides vary to achieve maximum sequence diversity in displays. Longer arbitrary primers are also useful and allow restriction sites to be engineered into the 5′-terminal end of the primer (e.g. AAGCTTGATTGCC with a 5′-terminal *Hind*III site) (Liang *et al.*, 1994). Arbitrary primers as long as 20 bases have been used in conjunction with extended oligo(dT) primers (Zhao *et al.*, 1995). These longer primers allow higher annealing temperatures to be used after initial rounds of degenerate priming, which may enhance the reproducibility of differential display. As with the oligo(dT) primers, a number of commercial firms offer collections of presynthesized arbitrary primers.

3.2.6 *Labelling of differential display products*

Initial reactions performed by Liang and Pardee utilized [α-^{35}S]dATP as the labelling agent for differential display. Because of its relatively low-energy β emissions, [α-^{35}S]dATP offers good band resolution upon autoradiography and presents low exposure hazard to researchers, a particular advantage if many reactions are being handled as part of an experiment. However, under routine thermocycling conditions, the [α-^{35}S]dATP label also forms a volatile decomposition product that penetrates the plastic amplification tubes (Liang and Pardee, 1995; Trentmann *et al.*, 1995). Although contaminated heat conducting fluids, such as oil or glycerol, can be removed from the thermocycler and the heating block can be cleaned with detergents, complete decontamination of the thermocycler can be difficult. Thus, if [α-^{35}S]dATP is used, it may be necessary to dedicate a thermocycler to differential display reactions and place it in a vented fume hood.

Several alternatives to [α-^{35}S] dATP labelling are available. With an energy emission spectrum and half-life intermediate between ^{35}S and ^{32}P, ^{33}P offers acceptable autoradiographic resolution and requires less extensive shielding than ^{32}P. [α-^{33}P] dATP is typically two- to three-fold more expensive than the corresponding ^{35}S- or ^{32}P-labelled nucleotide, but less nucleotide can be used in each reaction (Liang and Pardee, 1995). ^{32}P-labelled nucleotides offer a second labelling option. Exposure times are shorter and gels can be exposed without drying (Callard *et al.*, 1994), but the higher-energy β emissions result in significantly decreased resolution and increased radiation exposure. ^{33}P- or ^{32}P-labelled nucleotide decomposition products formed during thermocycling do not appear to penetrate the reaction tubes and contaminate the thermocycler. Finally, a non-radioactive method for visualization of differential display products by silver staining has been developed (Lohmann *et al.*, 1995).

3.2.7 *PCR and control reactions*

As for any technique based upon PCR, precautions must be taken to ensure reagents are free from contamination. Water and all solutions should be sterile and handled carefully to avoid introducing nucleases or nucleic acids. Even a trace amount of contaminating DNA can lead to the appearance of spurious bands in the display. Stocks such as 10× reaction buffer, nucleotides, primers and *Taq* polymerase should be stored in small aliquots at −20°C.

Consistency is an important element in successful execution of differential display. To reduce variation due to pipetting errors, prepare a master PCR mix for use in Protocol 3.3. For example, if only the first-strand cDNA template varies between reactions, prepare a mix including everything but the template. Components unique to specific reactions are added directly to the appropriate amplification tube. Keep the PCR mix and completed reactions on ice until ready to initiate the thermocycling programme. This will reduce both the *Taq* polymerase activity and the incidence of non-specific priming events.

Before investing time in an extensive search for differential products, several control amplifications should be done. First, an amplification using mock cDNA (from a reverse transcription reaction lacking enzyme) should be set up to test for the presence of contaminating DNA in RNA preparations and reagents. These products can be distinguished from those due to reverse transcription by *Taq* polymerase, by pre-incubating one set of reactions with DNase-free RNase. Second, several different oligo(dT) and arbitrary primer combinations should be tried with a single RNA sample. In the display autoradiograms, a large range of products extending to more than 500 bases in length should be observed. The display pattern should change when different primer sets are used, although some individual display products may be present in all the reactions using a given arbitrary primer. These products arise from the arbitrary primer annealing onto the template in opposite orientations (see control reactions in Figure 3.2, panel B, lane 6). The T6 product (Figure 3.3) is an example of this kind of product and was cloned from three different cDNA preparations amplified with the same arbitrary primer. Finally, the investigator should demonstrate that the patterns are reproducible by repeating the amplifications with freshly transcribed cDNA. As in Figure 3.2, panel C, the reactions should not vary significantly in terms of the size range or qualitative pattern of products.

3.2.8 *Electrophoresis*

Standard denaturing sequencing gels can be used to separate the differential display products. Samples should be separated by empty lanes (when using a sharkstooth comb) or the gel should be prepared using a well-forming comb so that bands can be excised without contamination from adjacent lanes. Only a portion of each amplification needs to be loaded onto the display gel; the remainder of each reaction can be stored at −20°C in the event that additional gels are needed. To achieve good resolution of larger products, electrophoresis should continue until products smaller than ∼100 bases are eluted from the gel. Larger products are generally easier to clone and label. In addition, larger products are more likely to include some coding sequence and therefore may be more useful for database comparisons.

PROTOCOL 3.3   PCR AMPLIFICATION AND GEL ELECTROPHORESIS

1. For each amplification reaction mix:

    2 µl first-strand cDNA template (1/10 of the first-strand reaction)

    2 µl of 10× PCR buffer

    2 µl 20 µM (each) dNTP mix

    2 µl 25 µM 3′ primer (e.g. T$_{12}$CG)

    0.5 µl 20 µM arbitrary 5′ primer

    1 µl [α-^{35}S] dATP (10 mCi/ml at 1200 Ci/mmol) or 0.25 µl of [α-^{33}P] dATP (10 mCi/ml at 1000–3000 Ci/mmol) plus 0.75 µl H$_2$O. End-labelled oligo(dT) primer may be substituted for free nucleotide, but this will bias the display pattern toward smaller amplification products (Tokuyama and Takeda, 1995).

    0.25 µl *Taq* polymerase (5 units/µl)

2. Overlay sample with mineral oil if necessary and amplify cDNA with the following cycling programme:

    94°C for 1 min

    [94°C for 30 s, 42°C for 1 min, 72°C for 30 s] × 40

    72°C for 5 min

    The reactions can be held at 4°C until removed and then stored at −20°C until gel loading.

3. Prepare a 6% denaturing polyacrylamide sequencing gel, containing 7 M urea, 1× TBE buffer, 6% (w/v) acrylamide (19:1 acrylamide:bis-acrylamide). Polymerize by addition of TEMED to 0.8 µl/ml and ammonium persulphate (from fresh 10% stock) to a final concentration of 0.4%.

4. Add 12 µl of denaturing/loading solution to each amplification reaction and heat for 2 min at 75°C. Load samples promptly into adjacent lanes of the sequencing gel. For a typical gel of 45 cm in length, the samples are electrophoresed at 45 W constant power for 2–3 h. The bromophenol blue dye will have the mobility of an approximately 20- to 30-base product, while the xylene cyanol FF will migrate as an approximately 160-base product. The loading of a lane of radiolabelled size standards (e.g. a ddG sequencing reaction of a known sequence) permits product size to be more accurately estimated when viewing the autoradiogram.

5. Dry gel at 80°C without fixation.

6. Place orientation markers on the dried gel prior to autoradiography. Applying small dots of phosphorescent acrylic paint with a pipette tip or needle allows the autoradiogram and gel to be aligned before excising bands for re-amplification. Usually an overnight exposure is sufficient to visualize the reaction products.

3.2.9 *Selection, recovery and cloning of amplification products*

Sequences that are likely to be differentially expressed can be identified by comparing the pattern of bands derived from different RNA samples. Because the amplification step does not sample the RNA populations in a highly quantitative fashion, qualitative differences (i.e. whether a particular band is present or absent) tend to be more reliable than quantitative differences (variations in band intensity). Some products, particularly species of lower molecular weight, will appear as clusters of bands. These generally represent sequences that differ in length by one nucleotide (due to presence or absence of a non-template nucleotide added by *Taq* polymerase) and/or differences in the migration rates of the separated strands. To reduce the possibility of obtaining additional contaminating sequences, choose only a single band for elution.

Selected products can be cloned after elution and re-amplification of the DNA in an excised band. Diffusion during boiling typically releases enough DNA to serve as template in an amplification reaction using the same primer set that generated the differential display product. The re-amplified product is then a stable source of DNA for a variety of purposes, including probe synthesis and cloning. Because *Taq* polymerase usually adds a non-template nucleotide, typically an adenine, to the $3'$ end of amplification products, they can be cloned directly into a vector prepared with a single thymidine overhang (T-A cloning) (Marchuck *et al.*, 1991). Commercial cloning vectors are available for this purpose but suitable vectors can be easily prepared by the investigator (see Protocol 3.4). If restriction sites have been incorporated into the oligo(dT) and arbitrary primers, more efficient multibase overhang ligations into a compatibly digested vector are possible. However, the ability of the chosen restriction enzyme to digest near the end of a DNA strand should be checked and an intermediate cloning step using a T-A overhang ligation may be necessary.

Nucleotide sequence data can be obtained from the cloned differential display product using oligonucleotide primers flanking the cloning site. Flanking primers must be used since arbitrary decamers and oligo(dT) primers (of the form $T_{12}VN$, where N = any nucleotide) are too short to be satisfactory sequencing primers. Differential display products that are produced with longer arbitrary primers of approximately 20 bases are more suitable candidates for direct sequencing because of the greater specificity of the longer primer.

PROTOCOL 3.4 RE-AMPLIFICATION AND CLONING OF DIFFERENTIAL AMPLIFICATION PRODUCTS

1. Place the autoradiogram behind the dried gel and view on a light box to see the band(s) through the attached blotting paper (fainter bands may need to be marked for accurate excision). With a clean blade, cut through the gel and blotting paper, removing the band with as little excess gel material as possible.

2. Place gel and backing paper into a tube and add 100 μl of distilled H$_2$O. Let gel slice rehydrate for 15 min and then boil for 15 min. Spin briefly in a microfuge, transfer the supernatant to a new tube and add 2.5 μl of 20 mg/ml nuclease-free glycogen, 10 μl 3 M sodium acetate pH 5.2, and 250 μl 100% ethanol. Precipitate for 30 min at −70°.

3. Pellet in a microfuge for 10 min at 16 000 *g*. Rinse pellet with 70% ethanol, dry, and resuspend in 10 μl of H$_2$O. Store at −20°.

4. Reamplify 2 μl of eluted DNA using the same primer combination, reaction conditions and amplification parameters used to produce the differential product, but increase final dNTP concentration to 20 μM (10-fold higher) and omit the radiolabel.

5. Electrophorese the amplification products on 1.5% agarose gel. Include size markers to check that products match the sizes observed on the display gel.

6. Excise band and electroelute or use a spin column to isolate the DNA from the agarose. Phenol:chloroform (1:1) extract and ethanol precipitate. Store at −20°.

7. Set up a 10 μl ligation reaction and incubate for several hours to overnight at 16°C.

   1 μl 10× buffer (300 mM Tris-Cl pH 7.8, 100 mM MgCl$_2$, 100 mM DTT, 5 mM ATP)

   100 ng of T-overhang vector[1]

   50–100 ng of isolated differential display product

   2 units T4 polynucleotide kinase

   5 units T4 DNA ligase

   H$_2$O to 10 μl

8. Transform into a bacterial host and select on appropriate media.

Notes

[1] Prepare a 100 μl PCR reaction containing 5 μg of plasmid digested with a restriction enzyme that generates a blunt end. Do not include primers and add dTTP to 100 μM final concentration as the only nucleotide. Add 5 units of *Taq* polymerase and incubate for 2 h at 72°C. A single non-template thymidine residue will be added to the 3′ ends of the plasmid. Extract the finished reaction with phenol:chloroform (1:1), ethanol precipitate and resuspend at 0.1 μg/μl. Store at −20°.

3.3 Troubleshooting

3.3.1 *Amplification products are small or absent*

Large amplification products cannot be generated from truncated cDNAs. Use the following protocol to check the quality of the first-strand cDNA. The resulting autoradiogram should reveal a continuous smear of synthesized cDNA with a significant amount of cDNA over 1 kb in length (see Figure 3.2, panel A).

PROTOCOL 3.5   cDNA SYNTHESIS TEST

1. Set up a first-strand reaction as given in Protocol 3.2. Immediately after adding reverse transcriptase solution, remove 5 μl to a separate tube containing 0.5 μl of [α-^{32}P]dCTP. Incubate the two reactions for 1 h at 37°C.

2. Add 50 μl of TE to the labelled reaction, pass the diluted reaction through a 1 ml column of Sephadex G-50 to remove unincorporated nucleotide and vacuum dry.

3. Resuspend the labelled cDNA pellet in alkaline loading buffer and load onto an alkaline agarose gel (1.5% agarose gel prepared in 50 mM NaCl, 1 mM EDTA. After the gel has set, immerse in 30 mM NaOH, 1 mM EDTA for approximately 30 min prior to loading – agarose will not gel properly in an alkaline environment). The running buffer is 30 mM NaOH, 1 mM EDTA.

4. Electrophorese with end-labelled DNA standards in adjacent lanes for size comparison and then dry gel (~45 min without heat to prevent melting of gel, then 30 min at 80°C). Expose dried gel to X-ray film.

Failure to obtain high molecular weight cDNA often stems from degradation of the RNA template prior to reverse transcription. Due to the durable and often ubiquitous nature of RNase in the environment, exogenous RNases are easily introduced inadvertently by laboratory personnel, microbial growth, or from equipment. Plant tissues themselves contain adequate quantities of RNase to degrade their cellular RNA. Thus, thawing of frozen material (and mixing of cellular contents by ice crystal damage) in the absence of protein denaturants (e.g. phenol) should be avoided. Care should be taken that gel rigs, combs, etc. that are also used for electrophoresis of DNA, particularly samples that contain RNase, are thoroughly cleaned before examining the integrity of an RNA sample. Incubation of the gel apparatus in 3% hydrogen peroxide for 15 min, followed by thorough rinsing with RNase-free water, can help to eliminate RNase contamination.

If strong distinct rRNA bands are not seen when total RNA is run on an agarose gel, the mRNA is likely to be significantly degraded as well. In that case, suspect solutions should be remade and stringent precautions taken to ensure maintenance of an RNase-free environment. Solutions should be prepared from RNase-free water, autoclaved, and maintained sterile. Autoclaving itself is not sufficient to destroy RNases. Glass-distilled or deionized water is usually sufficient for use, but may require treatment with the strong alkylating agent, diethyl pyrocarbonate (a hazardous reagent that should be handled carefully). Diethyl pyrocarbonate can be added to water (0.5 ml per litre), stirred for 20 min in a vented fume hood and then autoclaved for 30 min to 1 h. Sterile disposable plasticware, as supplied by the manufacturer, is generally free of RNase contamination. Glassware and metal utensils should be baked at 150°C for 3 or more hours. Establishing a supply of new reagents, glassware and plastics dedicated to use with RNA samples can help to maintain RNase-free conditions.

3.3.2 *Appearance of a continuous ladder of amplification products on the display gel*

The annealing temperature may be too low for the primer set in question. As this temperature is lowered, an increasing number of degenerate primer–template interactions become possible. Eventually, every position on the sequencing gel may be occupied by amplification products, making selection of differential products difficult or impossible. Raising the annealing temperature will favour primer–template combinations of increasing complementarity, reducing the number of observed products. Arbitrary decamer and oligo(dT) (T_{12}VN) primer sets are often used with annealing temperatures ranging from 37°C to 42°C. With longer primers, e.g. 20-mers, the first amplification cycle uses a 40°C annealing temperature for degenerate priming and is followed by amplification cycles with a 60°C annealing temperature (Zhao *et al.*, 1995). Primers contaminated with truncated oligonucleotides can also cause problems. Obtain oligonucleotide primers from sources that perform quality control analyses, including PAGE.

3.3.3 *Inability to amplify products from differential display gels*

The amount of DNA recovered by boiling an excised gel piece is quite low and may be a particular problem when dealing with high molecular weight species. Electroelution from a rehydrated gel slice is more effective and can boost recovery by close to an order of magnitude in some instances. Amplification with a larger quantity of template may yield enough re-amplified product to be visible on an agarose gel. In some cases, a second round of amplification using 1/100 of the first re-amplification may be necessary. However, extensive re-amplification (80 cycles in total) is not routinely recommended because the chance of amplifying contaminants increases greatly.

3.3.4 *Inability to recover recombinant plasmids with amplification products*

Preparation of the T-vector can be checked by setting up a control ligation and transformation with T-vector alone. The self-incompatible overhang should inhibit ligation and subsequently produce a very low number of transformants. Large numbers of transformants indicate incomplete digestion with the blunt-end restriction enzyme and/or incomplete thymidine addition. Gel purification of digested vector may be necessary to exclude trace quantities of uncut vector before addition of terminal thymidine residues.

3.3.5 *False positives*

False positives are defined as differential display products that exhibit a non-differential pattern of expression when subsequent analyses are performed for confirmatory purposes. Given the sensitivity of differential display to minor variations in technique and the fact that RNA populations are not sampled quantitatively, a certain level of false positives may be inherent in this method. The level of false positives can be reduced by attention to technique and inclusion of control reactions. Running duplicate samples to establish that a differential pattern is reproducible has been suggested as one useful countermeasure (Liang *et al.*, 1993). The most efficient approach would be to perform replications of this type only after candidate bands/sequences have been identified by a first round of displays. False positives may also arise from selection of contaminating sequences in the cloning step. Multiple amplification products can migrate with the same mobility on the display gel and will behave similarly upon agarose gel electrophoresis (Bauer *et al.*, 1993; Callard *et al.*,

1994). Selection of multiple cDNA clones after transformation is one answer to this problem. Several clones derived from a single differential band can be dot blotted onto duplicate membranes and screened with radiolabelled cDNA prepared from the poly[A]$^+$RNA populations being investigated. Clones that do not exhibit a differential signal can be excluded from further analyses. However, clones corresponding to rare messages will be difficult to assess by this method. In these instances, larger quantities of poly[A]$^+$RNA and Northern analyses may be required to test for differential expression.

Contamination by foreign organisms (see Section 3.2.1) is another source of false positives. Genomic Southern blots may be necessary to confirm that a differential display product originates from the organism under study. Southern analysis can also provide information on the genomic complexity of a given sequence. Such information can be useful in distinguishing between true and false positives in instances where the results of the Northern analyses differ substantially from the differential display patterns. For example, if the cloned product represents a differentially expressed member of a conserved and more widely expressed multigene family, its expression pattern could be obscured by cross-hybridization with mRNAs from related genes on Northern blots. In general, the more information the investigator can gather on a set of candidate clones, the easier it will be to separate false positives from sequences that represent truly differentially expressed genes.

3.4 Solutions

Alkaline loading buffer: 50 mM NaOH, 1 mM EDTA, 3% Ficoll 400, and 0.05% bromophenol blue.

Denaturing/loading solution: 95% formamide, 20 mM EDTA, 0.05% bromophenol blue, 0.05% xylene cyanol FF.

Homogenization buffer: 100 mM Tris-Cl pH 9, 100 mM NaCl, 1 mM EDTA, 0.5% SDS.

Loading dye: 20% Ficoll 400, 100 mM EDTA pH 8.0, 1% SDS, 0.25% bromophenol blue.

PCR buffer, 10×: 500 mM KCl, 100 mM Tris-Cl pH 8.3, 15 mM MgCl$_2$, 0.01% gelatin.

Phenol, buffer saturated: Add approximately one quarter volume of homogenization buffer to phenol previously equilibrated with Tris-Cl pH 8, and mix. Allow phases to separate. The phenol phase is usable when the layers are distinctly separated, although some cloudiness may remain.

Reverse transcriptase buffer, 5×: 250 mM Tris-Cl pH 8.3, 375 mM KCl, 15 mM MgCl$_2$.

TAE: 40 mM Tris-acetate pH 8.5, 2 mM EDTA.

TBE buffer, 10×: 162 g Tris base, 27.5 boric acid, 9.3 g Na$_2$EDTA in 1 litre H$_2$O.

Acknowledgements

The authors thank Dr Sharyn Perry and Talila Golan for helpful comments. Research in this laboratory has been supported by grants from the University of Wisconsin-Madison Graduate School; National Science Foundation (DCB-

9105527); Dept of Energy, National Science Foundation; Dept of Agriculture Collaborative Program on Research in Plant Biology (BIR-9220331); and National Institutes of Health Developmental Biology Training Program (5 T32 HD0718-14).

References

BAUER, D., MULLER, H., REICH, J., REIDEL, H., AHRENKIEL, V., WARTHOE, P. and STRAUSS, M., 1993, Identification of differentially expressed mRNA species by an improved display technique (DDRT-PCR), *Nucleic Acids Research*, **21**, 4272–4280.

CALLARD, D., LESCURE, B. and MAZZOLINI, L., 1994, A method for the elimination of false positives generated by the mRNA differential display technique, *Biotechniques*, **16**, 1096–1103.

FINKELSTEIN, R.R., TENBARGE, K.M., SHUMWAY, J.E. and CROUCH, M.L., 1985, Role of ABA in maturation of rapeseed embryos, *Plant Physiology*, **78**, 630–636.

HECK, G.R., PERRY, S.E., NICHOLS, K.W. and FERNANDEZ, D.E., 1995, AGL15, a MADS domain protein expressed in developing embryos, *Plant Cell*, **7**, 1271–1282.

JOHNSON, R.R., CRANSTON, H.J., CHAVERRA, M.E. and DYER, W.E., 1995, Characterization of cDNA clones for differentially expressed genes in embryos of dormant and nondormant *Avena fatua* L. caryopses, *Plant Molecular Biology*, **28**, 113–122.

LIANG, P. and PARDEE, A.B., 1992, Differential display of eukaryotic messenger RNA by means of the polymerase chain reaction, *Science*, **257**, 967–971.

1995, Alternatives to ^{35}S as a label for the differential display of eukaryotic messenger RNA, *Science*, **267**, 1186–1187.

LIANG, P., AVERBOUKH, L. and PARDEE, A.B., 1993, Distribution and cloning of eukaryotic mRNAs by means of differential display: refinements and optimisation, *Nucleic Acids Research*, **21**, 3269–3275.

LIANG, P., ZHU, W., ZHANG, X., GUO, Z., O'CONNELL, R.P., AVERBOUKH, L., WANG, F. and PARDEE, A.B., 1994, Differential display using one-base anchored oligo(dT) primers, *Nucleic Acids Research*, **22**, 5763–5764.

LOHMANN, J., SCHICKLE, H. and BOSCH, C.G., 1995, REN display, a rapid and efficient method for non-radioactive differential display and mRNA isolation, *Biotechniques*, **18**, 220–202.

MARCHUCK, D., DRUMM, D., SAULINO, A. and COLLINS, F.S., 1991, Construction of T-vectors, a rapid and general system for direct cloning of unmodified PCR products, *Nucleic Acids Research*, **19**, 1154.

TOKUYAMA, Y. and TAKEDA, J., 1995, Use of ^{33}P-labelled primer increases the sensitivity and specificity of mRNA differential display, *Biotechniques*, **18**, 424 426.

TRENTMANN, S.M., VAN DEr KNAAP, E., KENDE, H., LIANG, P. and PARDEE, A.B., 1995, Alternatives to ^{35}S as a label for the differential display of eukaryotic messenger RNA, *Science*, **267**, 1186.

VAN DER KNAAP, E. and KENDE, H., 1995, Identification of a gibberellin-induced gene in deepwater rice using differential display of mRNA, *Plant Molecular Biology*, **28**, 589–592.

WILKINSON, J.Q., LANAHAN, M.B., CONNER, T.W. and KLEE, H.J., 1995, Identification of mRNAs with enhanced expression in ripening strawberry fruit using polymerase chain reaction differential display, *Plant Molecular Biology*, **27**, 1097–1108.

ZHAO, S., OOI, S.L. and PARDEE, A.B., 1995, New primer strategy improves precision of differential display, *Biotechniques*, **18**, 842–850.

The Yeast Two-Hybrid System

SUSANNE E. KOHALMI, JACEK NOWAK, WILLIAM L. CROSBY

4.1 Introduction

The yeast two-hybrid system is a procedure for the genetic detection of protein–protein interactions in recombinant *Saccharomyces cerevisiae*. The system has been described in detail by others (Fields and Song, 1989, 1994; Chien *et al.*, 1991; Fritz and Green, 1992; Bartel *et al.*, 1993 a and b; Fields and Sternglanz, 1994; Phizicky and Fields, 1995). This chapter is intended to provide a practical account for the application of the two-hybrid system in a typical molecular biology laboratory setting. We have tried to emphasize experimental conditions which, in our experience, are helpful for minimizing the recovery of artefacts with this system. While our own practical experience has concentrated on the identification of protein-interaction partners in two-hybrid cDNA expression libraries of *Arabidopsis thaliana*, the protocols should prove generally useful for the construction and screening of libraries from other organisms.

4.1.2 *The GAL4-based two-hybrid system*

We use a variant of the yeast *GAL4*-based two-hybrid system originally described by Fields and Song (1989) and as modified by Chevray and Nathans (1992). In this version, the ability to detect protein–protein interactions is based on the unique properties of the gene product of *GAL4* as a positive transcriptional activator which co-ordinates the expression of genes required for galactose utilization in *S. cerevisiae* (Keegan *et al.*, 1986; Carey *et al.*, 1989). Two functional domains of GAL4 are important to the two-hybrid screening principle (Figure 4.1): the amino-terminal amino acids 1–147 constitute a DNA-binding (DB) domain which specifically recognizes a unique upstream activation sequence (UAS) in promoters of GAL4-regulated genes, while amino acids 768–881 encode a transcription activation (TA) domain which functions to recruit RNA polymerase II and activate transcription of nearby (downstream) genes. Importantly, transcriptional activation minimally requires the co-localization of both GAL4 domains at the promoter of a gene to be activated

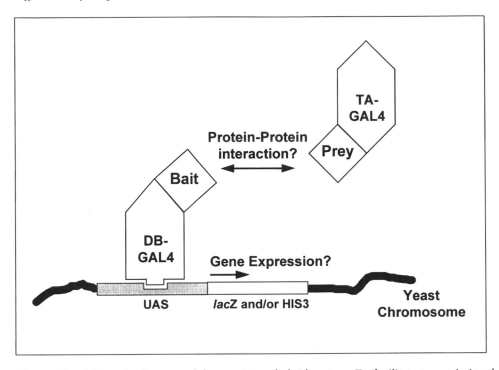

Figure 4.1 Schematic diagram of the yeast two-hybrid system. To facilitate transcriptional activation from GAL4 regulated genes only two of the GAL4-subdomains are required: the DNA binding (or DB) domain which recognizes a GAL4-specific UAS sequence in the promoter of GAL4-regulated genes and the transcription activation (or TA) domain which orchestrates the interaction with the transcription complex. A further prerequisite for the activation of a particular gene is that both the DB and TA domain are co-located at the promoter of the target gene. However, once disassembled, the two domains are unable to maintain an interaction on their own. Physical proximity can be re-established by fusing the GAL4-domains to two proteins which themselves have an interactive relationship. Only fusion proteins which are able to achieve such protein–protein interactions through their 'bait' and 'prey' components are able to activate transcription. For easy detection of GAL4- regulated gene expression, the GAL4-UAS sequence has been integrated into the promoter sequence of two reporter constructs (chromosomal integration) and gene expression can to be scored as resistance to 3-AT (*HIS3*) and/or as a change in visual appearance (*lacZ*).

(Figure 4.1). In the two-hybrid system, recombinant DB and TA domains of *GAL4* are expressed *in trans* from two independent fusion–expression plasmids that co-reside in the yeast host cell. In the absence of other factors, expression of the TA domain fails to activate transcription at UAS sites in the host since the promoter-localization function of the DB domain is absent. However, if the DB and TA domains are expressed as fusions to heterologous proteins which interact *in vivo*, then the co-localization of the DB and TA domains is reconstituted and activation of genes with UAS-containing promoters is realized.

Host cells for use in the two-hybrid system are deficient for the *GAL4* and *GAL80* genes normally required for galactose utilization. These strains contain reporter genes which harbour one or more UAS sequences incorporated into their respective promoters and integrated into the host chromosomes (Bartel *et al.*, 1993b). One reporter construct places the yeast *HIS3* gene under the control of *GAL4*, resulting in the expression of both histidine prototrophy and resistance to the toxic compound 3-amino-1′,2′,4′-triazole (3-AT). A second independent marker places the expression of *lacZ* (β-galactosidase) from *Escherichia coli* under the control of *GAL4*, permitting a convenient colorimetric assessment of marker gene activation. In these specialized host cells, marker gene activation is mediated by interacting GAL4-fusion proteins which act to reconstitute the physical proximity of DB and TA domains. Since marker gene activation results in resistance to 3-AT on media lacking histidine, the system is amenable to the forward genetic selection in host cells harbouring a known DB-protein fusion (the 'bait' construct) for rare protein interactions among cDNA–TA fusion libraries (the 'prey' construct). A high frequency of co-activation of the second *lacZ* reporter gene among 3-AT resistant colonies provides enhanced assurance that marker gene activation is the result of a bona fide protein interaction between the 'bait' and the product of an unknown cDNA.

The two-hybrid system presents several significant advantages: (i) technically, analysis of the structure–function basis of known interactions, as well as the analysis of novel protein interactions are facilitated by the straightforward characterization and manipulation at the DNA level afforded by the well-described yeast–*E. coli* shuttle plasmids employed in the system (Figure 4.2), (ii) the interaction potential of a particular 'bait' protein can be used to screen for new, unknown interactions *in*

1. Preparation of a known *GAL4*-DB fusion protein expression ('bait') construct
2. Assessment of the 'bait' construct in yeast (see below)
3. Preparation of a known 'prey' *GAL4*-TA fusion protein ('prey') construct or cDNA-TA fusion expression library
4. Assessment of the 'prey' construct against controls (not applicable to cDNA library constructs)
5. Co-transformation of 'bait' and 'prey' constructs into a 2-hybrid host strain (not applicable to library screening)
6. Screening for yeast colonies (resistant to 3-AT) expressing interacting fusion constructs
7. Examination of positive colonies for co-activation of the second (*lacZ*) reporter gene
8. Determination of cDNA insert size by yeast *in situ* PCR (optional)
9. Isolation of yeast DNA and transformation of shuttle expression plasmids into *E. coli* for further evaluation
10. Identification of *E. coli* transformants harbouring 'prey' constructs using TA construct-specific PCR or restriction analysis
11. Restriction enzyme characterization (fingerprinting) of cDNA inserts (optional)
12. Isolation of TA-construct plasmid DNA from appropriate *E. coli* colonies
13. Sequence characterization of the cDNA insert
14. Expression of the cDNA or other approaches leading to verification and characterization of the protein-protein interaction

Figure 4.2   Summary of the technical steps required for a two-hybrid screen.

vivo against large cDNA expression libraries, and (iii) the system is amenable to the study of known protein–protein interactions including the use of mutational strategies for the rapid dissection of structure–function relationships.

4.1.3 *Yeast strains*

Two different *S. cerevisiae* host strains are available for *GAL4*-based two-hybrid screening: YPB2 (Bartel *et al.*, 1993b), and HF7c (Feilotter *et al.*, 1994). The genotypes are described in Section 4.8. Both carry dual reporter genes for the detection of transcriptional activation (*lacZ* and *HIS3*) under the control of GAL4 UAS sequences in different promoter contexts. Two additional differences are worth noting between the two yeast strains: YPB2 harbours a mutation in the *CAN1* gene rendering this strain sensitive to the toxic arginine analogue canavanine, while HF7c contains a more stringently regulated *HIS3* reporter construct. In our experience both strains require the addition of 3-AT (albeit at different concentrations) for effective forward genetic screening described below. Consequently, HF7c may be used in place of, but presents no specific advantage over, strain YPB2.

4.1.4 DB *and* TA **fusion expression plasmids**

The two-hybrid expression plasmids in use in our laboratory have been designated pBI-880 and pBI-771, which are incremental modifications respectively of vectors pPC62 and pPC86 (Chevray and Nathans, 1992). Both carry multiple cloning site (MCS) for the cloning and expression of the 'bait' GAL4–DB fusion protein (pBI-880) or the GAL4–TA fusion protein (pBI-771), respectively. Both are yeast–*E. coli* shuttle vectors carrying yeast amino acid prototrophic markers (*LEU2* in pBI-880; *TRP1* in pBI-771), bacterial antibiotic resistance (*Amp*^{R}) for genetic selection, and replications origins (*col E1* for bacteria, *CEN6/ARS* for yeast) for propagation in both organisms. Since both the DB and TA vectors carry *Amp*^{R} markers, their distinction in *E. coli* is established by other means (see below). In both plasmids, expression of *GAL4*–DB or –TA fusions are under the control of the promoter and termination sequences from the yeast alcohol dehydrogenase I (*ADC1*) gene which confers constitutive expression (Bennetzen and Hall, 1982). Both expression plasmids incorporate N-terminal nuclear localization sequences to target the expressed fusion protein to the yeast nucleus (Chevray and Nathans, 1992). The plasmids pBI-880 and pBI-771 were modified from their pPC62 and pPC86 progenitors by the insertion of a FLAG epitope and enterokinase cleavage site upstream of the MCS, allowing the detection of fusion proteins using commercially available antibodies (e.g. M1 monoclonal antibody, IBI Kodak) and the cleavage of fused protein domains from their *GAL4* fusion partner (Figure 4.3 and 4.4).

4.1.5 *Growth conditions*

Yeast are grown on standard complete or minimal media as described by Sherman *et al.* (1983). Media for *E. coli* are prepared as described by Miller, 1972, Sambrook *et al.* (1989). Further details of media and other solutions are described in section 4.9.

```
DB-GAL4                                    FLAG/EK              SalI*
GTA TCG TCG AGG TCG AGG GAC TAC AAG GAC GAC GAC GAC AAG GGG TCG
 V   S   S   R   S   R   D   Y   K   D   D   D   D   K   G   S

       ClaI    HindIII EcoRV    EcoRI   PstI*   SmaI*   BamHI
ACG GTA TCG ATA AGC TTG ATA TCG AAT TCC TGC AGC CCG GGG GAT CCA
 T   V   S   I   S   L   I   S   N   S   C   S   P   G   D   P

SpeI*  XbaI*   NotI*            SacII*  T_ADC1
CTA GTT CTA GAG CGG CCG CCA CCG CGG TGG AGC TTT
 L   V   L   G   A   P   P   P   R   W   S   F
```

Figure 4.3 Nucleotide and deduced amino acid sequence of the Multiple Cloning Site (MCS) region of pBI-880, a modification of the vector pPC62, between the DB-*GAL4* domain and *ADC1* termination sequence. A FLAG epitope and enterokinase cleavage site (EK) have been inserted at the *Sal*I site. − partial yeast sequences (DB-*GAL4* and *ADC1* terminator); = FLAG/EK sequences; * unique restriction sites.

```
TA-GAL4                                    FLAG/EK              SalI
AAA GAG GGT GGG TCG AGG GAC TAC AAG GAC GAC GAC GAC AAG GGT CGA
 K   E   G   G   S   R   D   Y   K   D   D   D   D   K   G   R

SmaI    EcoRI    BglII   SpeI    NotI                        AatII
CCC CGG GAA TTC AGA TCT ACT AGT GCG GCC GCT AAG TAA GTA AGA CGT
 P   R   E   F   R   S   T   S   A   A   A   K   *   V   R   R

SacI                              SacII    T_ADC1
CGA GCT CTA AGT AAG TAA CGG CCG CCA CCG CGG TGG AGC TT
 R   A   L   S   K   *   R   P   P   P   R   W   S
```

Figure 4.4 Nucleotide and deduced amino acid sequence of the MCS region of the 'prey' vector pBI-771, a modification of the vector pPC86. The MCS is modified from pPC86 between the TA-*GAL4* and *ADC1* termination sequences by insertion of a FLAG epitope sequence and enterokinase cleavage site (EK). All restriction sites in the MCS are unique. − partial yeast sequences (TA-*GAL4* and *ADC1* terminator); = FLAG/EK sequences.

4.2 'Bait' and 'prey' vector constructs

4.2.1 Preparation of 'bait' and 'prey' vectors

In preparation for ligation, the 'bait' vector pBI-880 is restriction digested with *Sal*I + *Not*I and the resulting 8.3 kb vector fragment gel-purified from 0.8% agarose gel. The gene or protein domain of interest is subcloned or PCR amplified as the *Sal*I–*Not*I fragment defined by a 5′ [TCG] *Sal*I reading frame, and ligated to linearized pBI-880, resulting in an in frame fusion between DB-*GAL4* and the newly inserted gene (Figure 4.3). The 'prey' vector is similarly prepared except that the protein or domain of interest is ligated to pBI-771 linearized with *Sal*I and Not*I (Figure 4.4).

4.2.2 **Controls for the suitability of two-hybrid constructs**

Transform the 'prey' and 'bait' vectors into the yeast strain YPB2 (or HF7c) using a high-efficiency lithium acetate transformation protocol for yeast as described by Gietz *et al.* (1992) (or the Gietz web page: (http://www.umanitoba.ca/faculties/ medicine/human_genetics/gietz).

To confirm that specific 'bait' and 'prey' constructs are suitable for use in the two-hybrid screen, each construct is submitted to a series of controls as outlined in Table 4.1. None of the combinations described should result in the activation of reporter genes.

PROTOCOL 4.1 YEAST TRANSFORMATION PROTOCOL

For 1.5 ml of competent cells (about seven transformations) inoculate 300 ml of YPD, YPAD (for strains requiring adenine) or supplemented SD medium (for strains harbouring plasmids) from an overnight pre-culture such that growth overnight will yield a cell density of $5–10 \times 10^6$ cells/ml. Use the formula for inoculation density given in Figure 4.5. Note that YPB2 in SD medium reaches stationary phase at a cell density of approximately $2–2.5 \times 10^7$ cells/ml. To ensure that log phase cells are harvested, do not allow the culture to exceed 1×10^7 cells/ml, preferably harvesting the cells at about 7×10^6 cells/ml.

1. Transfer cells to sterile centrifuge bottle, centrifuge at 5000 rpm for 10 min in a large-volume rotor at room temperature. Discard the supernatant, resuspend the cells in 10 ml of glass-distilled sterile water and transfer to 40-ml sterile centrifuge tubes.

2. Centrifuge cells at 7000 rpm for 5 min in 50-ml tubes at room temperature. Resuspend the cells in 1.5 ml of Li acetate solution. If necessary, sonicate the cells for 3 min with an ultrasonicator (Bronson 2200) in an ice-cold water bath (sonication is only required with

---

Table 4.1 Required negative control transformations

Biological property avoided	Plasmid #1	Plasmid #2	Plating conditions (SD)
Transcriptional activation by the DB 'bait' construct	DB bait	–	-leu
Interaction between the 'bait' fusion protein and the TA-GAL4 domain	DB bait	TA	-leu-trp
Interaction between 'bait' fusion protein and an unrelated protein [1]	DB bait	TA–CRUCIFERIN fusion	-leu-trp
Transcriptional activation by the 'prey' fusion protein	–	TA prey	-trp
Interaction between the 'prey' fusion protein and the DB-GAL4 domain	DB	TA prey	-leu-trp
Interaction between the prey fusion protein and an unrelated protein [1]	DB–CRUCIFERIN fusion	TA prey	-leu-trp

[1]In our laboratory we use a full-length cDNA of the *Arabidopsis* seed storage protein CRUCIFERIN (Pang *et al.*, 1988) as a negative interaction control.

yeast strains which are clumpy or flocculent). Incubate for 1 h at 30°C with constant shaking.

3. Add transforming DNA to a sterile microfuge tube (for single transformants 1μl of a bacterial alkaline – SDS mini-prep yields sufficient transformants). Add up to 5 μg of transforming DNA in a volume not exceeding 20 μl and 20 μl of carrier DNA. Add 200 μl of competent yeast cells to each microfuge tube and mix gently. Incubate for 30 min at 30°C with shaking.

4. Resuspend the settled cells and add 1.2 ml of PEG solution; mix gently by inverting the tube and incubate for 30 min at 30°C with shaking. If the cells have settled to the bottom of the tube, mix gently by inversion. Heat shock for 15 min in a 42°C water bath then cool on ice.

5. Centrifuge microfuge tubes for 5 s and wash the cells twice with 0.5 ml sterile $1 \times$ TE buffer pH 8.0. Thoroughly but gently resuspend pelleted cells with the help of a sterile toothpick by twirling the flat-bottomed end in the solution.

6. Resuspend the cells in a final 1 ml of $1 \times$ TE buffer pH 8.0 and plate 0.2 ml ($2–4 \times 10^7$ cells) per 82-mm plate containing 28–30 ml of selection media; for large-scale library screening, plate 0.5 ml aliquots ($1–2 \times 10^8$ cells) to 145-mm plates containing 70–75 ml of selection media.

7. Incubate plates at 30°C until transformants appear; at 30°C this will take from 2–3 days when selecting for amino acid prototrophy, and up to 6–7 days on medium containing 3-AT.

For large-scale two-hybrid screens it is advisable to check the following: that the 'bait' construct-containing yeast strain cannot grow under selective conditions, the viable titre of the 'bait' yeast strain, and the titre of 'prey' construct-containing transformants.

To prepare stocks of frozen competent yeast cells, use the lithium acetate protocol described above. Following the 1 h incubation in Li acetate solution, transfer 200 μl of competent yeast cells to a sterile microfuge tube, add 30 μl of 100%

Assumed Parameters:

Growth Period: 17:00 to 10:00 = 17 h
Generation Time: 2 h
Number of Generations: 17 h ~ 8.5 generations
Cell Density of Overnight Pre-culture: 2×10^7 cells/ml

$$inoculation\ volume = \frac{desired\ cell\ density}{current\ cell\ density} \times \frac{culture\ volume}{2^{No.\ Generations}} = \frac{7 \times 10^6}{2 \times 10^7} \times \frac{300}{2^{8.5}} = 290\ \mu l$$

Note: If the overnight pre-culture is in stationary phase, you may need to add one additional generation to the No. of Generations value. By standardizing inoculation conditions (volumes and times), variables can be kept at a minimum and predicted cell densities become very reliable.

Figure 4.5 Formula for calculation of inoculation density using assumed parameters.

sterile glycerol, mix gently and freeze the cells to −70°C. Before use, transfer tubes to ice, thaw and add DNA. Transformation frequencies are generally lower when using frozen competent cells. For library transformation always use freshly prepared cells. However, for transforming single plasmids into yeast, or for simultaneous transformation of two different plasmids, frozen competent cells are adequate.

For convenient scale-up, all volumes may be doubled and 2.0-ml microfuge tubes used (note: PEG solution is added to fill the 2-ml tube). In our experience, this reduction in PEG volumes has no negative effects on transformation frequencies. Cells are finally resuspended to 1 ml, allowing for the plating of 0.25 ml aliquots (about 1×10^8 total cells) to each of four 145-mm screening plates. To achieve the highest transformation frequencies possible it is recommended to prepare competent cells from yeast already containing the 'bait' vector and use only aliquots of the cDNA library as transforming DNA.

4.2.3 *Transcriptional activation of reporter genes*

To determine if the expression of a single or combination of DB + TA fusion constructs is able to activate transcription, activation of both the *HIS3* and *lacZ* reporter constructs should be confirmed. For economy of effort, we simply re-spot two individual colonies representing a particular plasmid combination onto a master selection plate and grow for 1–2 days. The master is replica-plated onto appropriately supplemented [SD-His + 3-AT] media, and imprinted to a nitrocellulose filter to perform an Xgal filter assay, modified from Chevray and Nathans (1992). Interactive 'bait' and 'prey' expression constructs which activate both marker genes are scored as growth on 3-AT plates and blue-coloured cells in the Xgal filter assay.

PROTOCOL 4.2 XGAL COLONY FILTER ASSAY

Pure nitrocellulose filters are extremely brittle when submerged in liquid nitrogen, therefore use nylon-backed nitrocellulose filter disks (e.g. MSI 'NitroPlus').

1. Mark orientation on filters with pencil or ball point pen.

2. Place filter onto colonies or spots to replicate pattern. Allow filter to wet completely and avoid air bubbles. Lift filter carefully to avoid smearing.

3. Submerge or float the filter colony-side down for 5–10 s in liquid nitrogen to permeabilize cells, then carefully remove and place cell-side up in a Petri dish that contains filter paper (single layer) saturated with 1.8 ml of Z-buffer. Ensure that there are no air bubbles between filter and filter paper.

4. Incubate at 30°C for 10–60 min to overnight, or even up to several days.

Incubation time depends on the degree of *lacZ* marker activation. Strong activation in a *CEN/ARS* plasmid usually turn blue within 20 min, while weaker responses require more careful interpretation. Positive clones can be picked directly from the filter or the replica master and streaked immediately onto appropriate medium. A long-term record can be generated by soaking the nitrocellulose filter in 0.2 M NaHCO$_3$ before drying and storage.

4.2.4 *Analysis of GAL4 fusion proteins*

To confirm that the 'bait' of interest is expressed as an in-frame GAL4 fusion protein, two different approaches should be taken. Sequence the *GAL4*-FLAG-cDNA fusion junction to confirm the correct alignment. Directly assess, by Western blot analysis (Burnette, 1981), the predicted size of the 'bait' fusion protein (including an approximately 16 kDa contribution by the $GAL4_{1-147}$ domain) by probing recombinant yeast protein extracts using anti-GAL4, FLAG or protein-specific antibodies. In our hands, detection of different GAL4 fusion proteins in total yeast homogenates yields variable results in Western blots probed with high-titre anti-GAL4 polyclonal antibodies, probably due to variations in steady-state expression of the fusion protein. Yeast extracts prepared from *GAL4* deletion strains also express a number of endogenous proteins that cross-react with GAL4 antibodies, which can obscure the recombinant fusion protein signal of interest. Monoclonal antibodies against *GAL4* are now available from commercial sources (e.g. Santa Cruz Biochem, Clontech), which may improve the sensitivity and specificity of Western blots. For the detection of fusion proteins, we use any of several, commercially available, sensitive, chemiluminescent detection systems (e.g. the 'ECL' system, Amersham).

PROTOCOL 4.3   PREPARATION OF YEAST PROTEIN EXTRACTS

1. Inoculate 200 ml of appropriately supplemented SD medium from an overnight pre-culture so that growth overnight will yield a cell density of $5–10 \times 10^6$ cells/ml (use the calculation described for yeast transformation).

2. Transfer cells to sterile bottle(s) and centrifuge at 5000 rpm for 5 min in a GSA (Sorvall) or equivalent rotor. Discard the supernatant, resuspend the cells in the remaining liquid (approximately 1 ml) and transfer to a single, flat-bottomed screw cap, microfuge tube equipped with an O-ring.

3. Re-centrifuge and carefully aspirate away the remaining supernatant. Add 40 µl of Laemmli buffer and 400 µl of acid washed glass beads. To break the cells, vortex for 2×1 min, making sure that the glass beads are moving in the tube. If there is not enough liquid the beads usually do not move at all, whereas too much liquid will result in the glass beads remaining settled and only the cells being vortexed. In both cases breakage of the cells will be suboptimal. Breakage efficiency can be checked by light microscopy.

4. Add 960 µl of Laemmli buffer and transfer tube immediately to boiling water. Boil for 3 min and cool on ice. Microfuge for 5 min at 4°C and transfer the supernatant to a fresh microfuge tube. Store the protein extract at -20°C.

For Western blots, 1 µl of above extract is sufficient to realize a clear signal following a 20 s exposure to the ECL detection solutions.

4.3 Preparation of a two-hybrid cDNA expression library

Prior to the construction of any cDNA library there are a number of factors to consider. One is the source of tissue for poly[A]$^+$ mRNA preparation used as the template for cDNA synthesis. Organ- or tissue-specific libraries are attractive if it is anticipated these tissues will be enriched for specific mRNAs of interest. A second factor is the priming strategy for the synthesis of first-strand cDNA by reverse transcriptase. The use of dT-containing oligomers will preferentially prime DNA synthesis from poly[A]$^+$ mRNA templates, but will result in a population of first-strand DNAs (and a library) somewhat overrepresentative of the transcript 3$'$ sequences. Random oligomers are less selective in priming poly[A]$^+$ compared with ribosomal RNA, although the resulting population of cDNA product will generally be more representative of all transcript sequence ends. Both approaches are amenable to directional cloning, which reduces by one-half the frequency of out-of-frame cDNAs, as all sequences are cloned into expression vectors in the 'sense' orientation. A third consideration is the number of independent cDNA clones, or diversity of the library itself. This may be an important consideration, especially for 'general' libraries constructed from poly[A]$^+$ mRNA isolated from pooled tissues where the frequency of individual cDNAs in the library may be low.

In our studies involving *Arabidopsis thaliana*, we used an oligo-dT primed library that would be of broad utility in subsequent two-hybrid screening. Whole plant tissues from *Arabidopsis* were collected at various growth stages from 1- to 4-week-old seedlings. Poly[A]$^+$ mRNA was isolated and cDNA prepared using commercial systems. cDNAs were directionally ligated as *Sal*I – *Not*I fragments into the vector pBI-771 and transformed into *E. coli* by electroporation. Transformants were pooled as primary cell stocks and stored in 7% (v:v) DMSO at $-80°C$. The resulting cDNA library contained 2×10^7 independent clones and was about 95% recombinant by direct plasmid analysis. Aliquots of the cell stocks (containing at least 10^9 viable cells) were cultured in 5 litres of standard selection medium and large-volume plasmid DNA preparations were carried out using a commercial kit ('Giga Kit'; Qiagen). Plasmid DNA (5–15 mg) was lyophilized and stored in 50–250 µg aliquots at $-20°C$. Prior to use, cDNA library DNA was reconstituted in sterile water at a final concentration of 1 mg/ml.

4.3.1 Screening two-hybrid expression libraries

4.3.1.1 Transforming with cDNA library DNA

To recap, freshly competent cells are prepared of yeast strain YPB2 (or HF7c) carrying the 'bait' vector, they are transformed with the TA-*GAL4*-cDNA fusion expression library in vector pBI-771 (Figure 4.4) and plated to [SD-Trp-Leu-His + 3-AT] medium. As controls, dilution aliquots are plated to [SD-Leu-Trp] and [SD-Leu] media to determine the transformed- and viable-cell titres, respectively.

A principle objective for library transformations is to recover the maximum possible number of transformants. Although yeast electroporation protocols are very attractive in terms of transformants per µg of plasmid DNA, the uptake procedure becomes saturated at relatively low concentrations of DNA resulting in the recovery of low absolute numbers of transformants. The Li acetate protocol (Gietz *et al.*, 1992) is simple, gives linear recoveries of transformants and allows for the generation of large numbers of competent cells that are required for extensive two-hybrid screens. We find approximately 0.1–0.2% of the viable cells become transformed, and 10^7 total transformants can be generated from 4 litres of culture in a single experiment. The number of transformants that constitute a complete screen, of course, depends in part on the diversity of the two-hybrid expression library being transformed. An exhaustive screen of a library requires a 3–10-fold higher frequency of transformants to be analysed relative to diversity of the library (Sambrook *et al.*, 1989). A second consideration is the abundance of positive-interacting cDNA products in the library, bearing in mind that they may be a subset of all cDNA clones representing a given transcript if only particular domains of the *GAL4*-TA fusion product (as partial cDNA clones) are competent for protein interaction with the bait.

In plants, a typical 'housekeeping' transcript (e.g. the branch-chain amino acid, anabolic enzyme acetolactate synthase) is expressed in the range of 10^{-5} relative abundance in the poly$[A]^+$ mRNA population (Bekkaoui *et al.*, 1993), while genes encoding regulatory factors may be considerably below this. In independent experiments, we have recovered protein–protein interactors with frequencies ranging from about 8×10^{-4} to 8×10^{-6} using a number of different 'bait' constructs (Kohalmi and Crosby, unpublished data). For initial interactor-'fishing' where one cannot anticipate a specific recovery frequency, we recommend small-scale experiments in order to reasonably assess the frequency of positive interactions that will be recovered. The recovery of one or a few bona fide positives from such experiments justifies a second large-scale screening. We often recover sufficient interacting clones (20–50) from such limited experiments.

In principle, competent yeast cells can be co-transformed with both 'bait' and cDNA library plasmid DNAs. However, the frequency of co-transformation with two independent vectors is significantly lower than the single event. Consequently, to achieve high numbers of 'bait' and library cDNA plasmids in the same yeast cell, competent cells are prepared from yeast already carrying the 'bait' expression plasmid, and subsequently singularly transformed with cDNA library DNA.

4.3.1.2 Hybrid screening conditions

We are aware of a considerable oral tradition regarding the tendency of the two-hybrid screen to yield artefactual protein interactions. We have experimental data to suggest that such artefacts can be largely avoided by the use of appropriate plating conditions during the screen itself. While artefacts can be readily recovered under suboptimal conditions, they are not an inherent property of the two-hybrid system.

To determine appropriate plating conditions, we undertook a series of experiments beginning with transforming the host yeast strain (YPB2) with a single-copy *CEN-ARS* plasmid expressing a full-length *GAL4* gene product, which fully activated both two-hybrid marker genes. These cells were progressively diluted into a population of the same cells carrying a non-recombinant plasmid. Aliquots of the dilution series were plated at various cell densities onto media containing different 3-AT concentrations, and the number of 3-AT-resistant colonies scored as a fraction of the total cells plated. Artefacts were classified as those colonies which were phenotypically 3-AT resistant, but which did not activate the *lacZ* marker and did not contain the *GAL4*-expressing plasmid construct. These reconstruction tests indicated that elevated plating densities contributed to an increased frequency of artefacts in the population. For strain YPB2, optimal plating conditions were defined as approximately 2×10^7 viable cells per standard 82-mm Petri dish on SD medium containing 5 mM 3-AT. Under these conditions we routinely encounter less than 5% of 3-AT-resistant colonies which do not co-activate the *lacZ* marker. To date, our experiments have indicated a high frequency of protein interactions which are intuitively reasonable for the particular 'bait' construct under study. In some experiments involving selected bait constructs, extensive screening of the library failed to present a single positive (3-AT resistant) colony. While such results may be disappointing, they are reassuring in the sense that the system as employed does not offer a high frequency of 3-AT-resistant colonies as artefacts.

The absolute 3-AT concentrations appropriate for a two-hybrid screening experiment will vary depending on the yeast strain used for the experiment. Furthermore, the use of single-copy or multi-copy vectors could influence steady-state expression levels of interacting fusion protein complexes, which can in turn bias the 3-AT concentration required to score a positive interaction reliably.

The optimal plating conditions described here should not imply that 3-AT-containing plates will not exhibit background growth. We have opted to use the minimal 3-AT concentration required to distinguish between positive cells activating the marker genes, from the background of non-activating microcolonies. While 5 mM 3-AT does not provide for a highly stringent suppression of background growth, it does allow for the unambiguous identification of positive colonies in strain YPB2. Incubated under these selection conditions, a typical screening plate will exhibit a background containing numerous tiny (needle-tip sized) colonies visible at a frequency one would expect for the total number of transformed cells. In addition, per 145-mm plate one finds approximately 2 to 10 small, flat colonies (0.5–1 mm in diameter) which generally do not co-activate *lac*Z. Finally and most rewarding, positives are occasionally evident as very distinct, large, hill-shaped colonies 2–4 mm in diameter, over 95% of which elicit a significant *lac*Z signal in filter assays.

4.4 Processing and characterization of 3-AT resistant yeast colonies

Before processing 3-AT-resistant colonies for the recovery of plasmid DNA, record the diameter of positive colonies. Among plates growing under identical conditions, we find that the diameter of the colonies is often a useful indicator of interactor-dependent gene activation in subsequent *lac*Z assays.

4.4.1 *Processing of 3-AT-resistant colonies*

For rapid processing, a single colony lifted with a toothpick can be used to generate enough cell material for freezing, to create a master plate for reporter gene analysis and to inoculate a culture for the subsequent isolation of plasmid DNA. At a minimum one has to establish safely archived −80°C stocks which can be conveniently processed at a later time. To freeze a sample of each colony, simply re-streak cells on a fresh plate to generate enough material for archiving. Pick up cells using a sterile toothpick from the centre of a single colony, avoiding transfer of any background growth. Small colonies should be re-streaked on selective media containing 3-AT to reduce contamination with background cells. After 2–3 days of incubation, scrape cells, resuspend in 50% glycerol and archive at −80°C.

To test for the transcriptional activation of the *lac*Z reporter gene, use a sterile toothpick to transfer cells to a fresh plate; up to 50 colonies can be placed on a single plate and processed simultaneously. Check for *lac*Z expression using the Xgal filter assay as described above. The characteristic of one class of false positives is their ability to activate expression of only one of the reporter genes.

To isolate yeast DNA from 3-AT-resistant and *lacZ* positive colonies, inoculate 5 ml cultures of [SD-Trp] medium. Since both 'bait' and 'prey' vectors carry the same bacterial selective marker, statistically only half of the *E. coli* cells transformed with DNA isolated from positive yeast clones will harbour the desired cDNA sequence (the 'prey' vector). By growing colonies in medium selecting for the 'prey' vector ([SD-Trp] for pBI-771), one can skew the recovery in favour of those retaining the 'prey' vector.

PROTOCOL 4.4 RECOVERY OF PLASMID CONSTRUCTS FROM YEAST

This procedure is a rapid SDS-alkaline lysis procedure for preparation of yeast DNA modified from a protocol submitted to the yeast Internet forum by Robin Wright, Washington University, St. Louis. It is a 'quick and dirty' way to extract yeast DNA but the yields are adequate for further processing.

1. Inoculate a 5 ml yeast cell pre-culture and grow to stationary phase in [SD-Trp] medium (for pPC86 or pBI-771) for 1–2 days, selecting for the cDNA-containing 'prey' expression plasmid.

2. Centrifuge the cells at 7000 rpm in a Sorvall GSA rotor for 5 min, discard the supernatant and resuspend in 1 ml of zymolyase solution. Transfer the suspension to 1.5-ml micro-fuge tube and incubate at 37°C for 1 h. Check for spheroplasts by mixing 10 μl of cell suspension with 10 μl of the cell lysis solution. Observe by microscopy, comparing with 10 μl of untreated cell suspension. If more than 60% of the cells are lysed then proceed; if not, incubate for a further 15 min (treatment time is strain dependent) and recheck.

3. Microcentrifuge the cells for 5 to 15 s, aspirate the supernatant carefully and add 200 μl of cell lysis solution to the cell suspension. Thoroughly mix by pulling up and down through a small pipette tip and incubate at 65°C for 5 min.

4. Add 200 μl of neutralization solution. Mix gently by inversion, chill on ice for 15 min, then microcentrifuge at top speed for 20 min at 4°C.

5. Transfer the clear supernatant to a fresh 1.5-ml microfuge tube, add 1 ml of ice-cold 95% ethanol and mix by inversion. Spin for 2 min, aspirate dry and resuspend the pellet in 40 μl of water.

For yeast transformation use 5–10 μl of the DNA, and for electroporation of *E. coli* use 1–2 μl of DNA.

4.4.2 *Preparation of electrocompetent* E. coli *cells*

The above protocol for preparation of yeast DNA is relatively crude but it results in sufficient yields for *E. coli* transformations. Electroporations are recommended over CaCl$_2$ transformation protocols due to the much higher transformation efficiency. Although competent cells are commercially available, the processing of large number of 3-AT-resistant colonies makes it more economical to use 'homemade' electrocompetent cells as described in Chapter 2, Protocol 2.10.

PROTOCOL 4.5   ELECTROPORATION OF *E. coli* WITH YEAST DNA

Leave fresh cells on ice for 1–3 h or thaw frozen cells on ice.

1.   During this incubation time, transfer aliquots of yeast DNA into electroporation cuvettes on ice and for each transformation prepare 1 ml of 2x YT in microfuge tubes.

2.   Electroporate 100 µl of cells with 2 µl of yeast DNA. Use electroporation settings as described by the manufacturer or as in Chapter 2, Protocol 2.11.

3.   Transfer electroporated cells to 2x YT and incubate for 30–60 min at 37°C.

4.   Plate 0.2 ml of electroporated cells onto plates containing 2x YT + amp and incubate overnight (11–13 h) at 37°C.

4.4.3   *Identification of 'prey' vectors in* E. coli *transformants*

Both 'bait' and 'prey' vectors described here carry the same Amp^R marker for selection in *E. coli*. To identify colonies harbouring the 'prey' vector, *E. coli* transformants can be easily screened by direct PCR using primers specific for the *GAL4* trans-activation domain and the *ADH*1 termination sequence. Normally, over one-half of the transformed *E. coli* colonies should be positive; in our experience, screening two colonies per transformation event has an approximately 80% chance of recovering a 'prey' vector as evidenced by the amplification of a PCR product. As a control, one can compare the size of the recovered PCR product to those amplified directly from the progenitor yeast colony.

4.4.4   **In situ *PCR of cDNA segments***

4.4.4.1   *Preparation of yeast cells*

Yeast cells are frozen in 50% glycerol and stored at −80°C. For PCR, add 5 µl of a 1:10 dilution of cells in H_2O to the PCR reaction mix. We are much less successful with *in situ* PCR reactions when using fresh yeast cells. Also, addition of 20 mM NaOH to the reaction mix results in larger yields and more clearly defined PCR products from intact yeast cells (Wang *et al.*, 1996). The yield of yeast PCR is generally lower than *E. coli in situ* reactions, so we prefer to amplify the PCR products from bacterial cells for further processing. To visualize the PCR products, the entire 25 µl reaction is resolved on agarose gels.

4.4.4.2   *Preparation of* E. coli *cells*

Touch individual colonies with a sterile toothpick, transfer to 50 µl sterile water and transfer 5 µl to the PCR reaction mix. Generally 10 µl of a 25 µl standard PCR reaction are sufficient to visualize amplified products on an agarose gel. The remaining reaction can be used for further characterization of the cDNA insert (see below). To rescue a PCR-positive colony, remove cells for PCR reaction and immediately add 950 µl of 2 × YT + Amp to the remaining cell suspension, regrow and archive the cells to −80°C.

PROTOCOL 4.6 PCR

1. For each sample add:

    5 µl DNA sample

    2.5 µl 10 × PCR buffer

    5 µl 1 mM dNTP mix

    0.25 µl primer 1

    0.25 µl primer 2

    0.125 µl 5 units/µl *Taq* polymerase

    11.875 µl water.

2. The PCR cycle conditions are:

    96°C for 2 min

    [96°C for 45 s, 60°C for 1 min, 75°C for 2 min] × 30 cycles

    75°C for 10 min

    Hold at 15°C

For a 'prey' specific primer we use BC304 [5'-CTATTCGATGATGAAGATACC-3']; for an ADH1-terminator primer we use JN069 [5'-TTGATTGGAGACTTGACC-3']. Primers are used at 100 pmol/µl.

4.5 Characterization of positive *E. coli* transformants

4.5.1 *DNA sequencing*

To sequence cDNA inserts, isolate plasmid DNA (Sambrook *et al.*, 1989) from at least one positive *E. coli* transformant representing each independent cDNA class. To avoid extensive sequencing of identical clones, precharacterisation of PCR-amplified DNA fragments by restriction endonuclease 'fingerprinting' has proven useful for a preliminary definition of cDNA classes. Portions of the PCR reaction (10–15 µl) can be cleaved with four or five-base site restriction enzymes. The restriction pattern is resolved on 4% polyacrylamide gels in 1× TBE and clones tentatively classified according to their restriction band pattern.

4.5.2 *Control transformations*

To ensure further that positive interactions in yeast host cells can be ascribed to the plasmids they contain, it is important to re-transform isolated 'prey' plasmids into yeast harbouring the 'bait' construct. Additional plasmid control combinations should exhibit a diagnostic behaviour as summarized in the Table 4.2.

These controls ensure that the protein interaction is specific for the cloned cDNA construct. The first three control transformations, when negative, confirm that the recovered cDNA expression construct does not itself activate transcription, and that the protein interaction is not mediated by the *GAL4*-DB domain of the 'bait' fusion protein. Reproducibility of a positive signal in the control is also important to provide direct confirmation that the recovered 'prey' plasmid correctly encodes an interacting protein.

Table 4.2 Control transformations

Assessment criteria	Plasmid #1	Plasmid #2	Plating media (SD)	Expected interaction capability
Transcriptional activation by fusion construct	–	TA-interacting-prey	-Trp	Negative
Interaction between fusion construct and corresponding *GAL4* domain	DB	TA-interacting-prey	-Leu-Trp	Negative
Interaction between fusion construct and unrelated protein[1]	DB–CRUCIFERIN	TA-interacting-prey	-Leu-Trp	Negative
Confirming specificity of interaction	DB-interacting-bait	TA-interacting-prey	-Leu-Trp	Positive

[1] see Table 4.1.

4.6 Potential problems

'Bait' constructs which activate transcription of the two-host reporter genes are unsuitable for use in a two-hybrid screen. We have successfully used modified 'bait' fusion constructs, partial domains of the 'bait' protein which are non-functional for trans-activation in two-hybrid screens. However, modification of the 'bait' protein may destroy potential protein–protein interaction sites, or cause subtle alterations of the overall three-dimensional structure of the GAL4 fusion protein so as to alter the specificity of the protein–protein interaction. Activation of transcription by a 'prey' construct is generally not a concern as long as the DNA-binding property of the interacting fusion proteins is determined by the *GAL4*-DB domain. Appropriate control transformations described above detect those cDNA-fusion constructs which bind to the promoter sequence. The use of yeast host strains carrying two reporter genes under the control of *GAL4* UAS in different promoter sequence contexts reduces the possibility of including this type of false positive, as it is unlikely that a specific 'prey' construct would recognize both independent promoters.

Clearly, failure of the two-hybrid approach to identify protein interactions involving a given bait construct does not exclude the possibility that they exist. The limitations of yeast as a heterologous expression host, along with biological requirements of the two-hybrid system, may combine to frustrate the recovery of positives. For example, constructs that are incompetent for transport to the yeast nucleus and effecting marker gene activation may not be recovered. Yeast hosts may lack specific factors which direct the correct folding or other post-translational modifications of a particular 'bait' protein that are required for it to present a functional protein-interacting domain. Additionally, the host may express endogenous proteins which interfere with the interaction of specific 'bait' and 'prey' constructs, which is a consideration for constructs involving evolutionarily highly conserved proteins. One can also anticipate that some 'prey' protein domains would not be competent for protein interaction as the C-terminal fusion partner with GAL4. Such an eventuality would require the development of N-terminal fusion expression vectors and/or the construction of N-terminal *GAL4*-TA cDNA expression libraries, but to our knowledge no such constructs or libraries are currently available. At present, the two-hybrid system is designed for the detection of binary protein interactions. 'Bait' proteins that require three or more obligatory participants to form a protein complex cannot readily be detected without modification of the system.

4.7 Outlook

As stated by Phizicky and Fields (1995), the yeast two-hybrid system offers a complementary approach for the analysis of protein interactions. As described above, the particular requirements of a yeast heterologous expression system may preclude the identification of interactions involving some 'bait' constructs. Since the potential always exists for the detection of artefactual interactions identified using the two-hybrid system, alternative approaches are important to confirm the existence and specificity of protein interactions. All approaches, including the yeast two-hybrid system constitute a starting point for understanding structural or mechanistic aspects of function for the protein of interest, including establishing the relevance of function in the source organism.

The two-hybrid system offers significant advantages by enabling the rapid genetic selection and analysis of unknown, potentially rare, protein interactions from among large cDNA expression repertoires. There are also indications that the system can identify protein interactions that are difficult to detect or analyse in the source organism for reasons of low interaction stability or low abundance of the interacting proteins (Phizicky and Fields, 1995).

The development of improved protocols, combined with ongoing development of the system and its associated technologies are likely to improve further the power and utility of the approach. These developments include the enhanced availability of private or commercially available, high-diversity, two-hybrid expression libraries from different sources. Recent developments include dedicated His$_6$-tagged expression vector systems for the purification of cDNA gene products recovered from two-hybrid libraries (Nowak, Schorr and Crosby, unpublished), and the development of 'reverse two-hybrid' systems for analysing the structure–function basis of specific protein–protein interactions (Vidal *et al.*, 1996a, 1996b).

4.8 Yeast strain genotypes

YPB2 is: [*MATa ura3-52 his3-200 ade2-101 lys2-801 trp1-901 leu2-3,112 can*^{R} *GAL4-542   gal80-538   LYS2::GAL1*$_{UAS}$*LEU2*$_{TATA}$HIS3   URA3::*GAL4* 17-mers($\times$3) *CyC1*$_{TATA}$*lacZ*] (Bartel *et al.*, 1993b).

HF7c is: [*MATa ura3-52 his3-200 ade2-101 lys2-801 trp1-901 leu2-3,112 GAL4-542 gal80-538 LYS2::GAL1*$_{UAS}$ *GAL1*$_{TATA}$ *HIS3 URA3::GAL4* 17-mers($\times$3) *CyC1*$_{TATA}$ *lacZ*] (Feilotter *et al.*, 1994).

4.9 Growth media, solutions and supplies

4.9.1 *Growth media*

Yeast peptone dextrose (YPD) medium: YPD is a 'complete' medium, that is, a complex mixture of protein and yeast hydrolysates which does not permit the prototrophic selection of plasmids. Mix 10 g yeast extract, 20 g bactopeptone, 20 g glucose, and water to 1 litre.

Synthetic dextrose (SD) medium: selective medium is prepared by supplementing a basic salt medium (yeast nitrogen base) with the appropriate 'dropout powder' – a mixture of amino acids plus adenine and uracil for selection of the desired prototrophy in yeast. Mix 20.0 g glucose, 6.7 g yeast nitrogen base without amino acids, 1.5 g appropriate dropout powder, and water to 1 litre.

Yeast nitrogen base without amino acids: this expensive component can be prepared by adventurous investigators with spare time and a ball-mill, at a fraction of the commercial price. Mix 5.0 g $(NH_4)_2SO_4$, 1.0 g KH_2PO_4, 500 mg $MgSO_4$. $7H_2O$, 100 mg NaCl, 100 mg $CaCl_2.2 H_2O$, 2 mg inositol, 500 µg H_3BO_3, 400 µg $ZnSO_4.7H_2O$, 400 µg $MnSO_4.4H_2O$, 400 µg thiamine HCl, 400 µg pyridoxine HCl, 400 µg niacin or nicotinic acid, 400 µg calcium pantothenate, 200 µg *p*-aminobenzoic acid, 200 µg riboflavin, 200 µg $FeCl_3$, 200 µg Na_2MoO_4. $4H_2O$, 100 µg KI, 40.0 µg $CuSO_4.5H_2O$, 2.0 µg folic acid, 2.0 µg biotin, and water to 1 litre. The final product pH is 5.5.

Amino acid mixes: filter-sterilized liquid stock solutions of all components can be prepared individually and aliquots can be added to media in any desired combination prior to autoclaving. For different selection purposes omit the appropriate component from the mix, e.g. the [-leu-trp] mix contains all the above components except for leucine and tryptophan. For each litre of medium mix 39.84 mg adenine sulphate, 19.92 mg L-arginine-HCl, 99.6 mg L-aspartic acid, 99.6 mg L-glutamic acid, 19.92 mg L-histidine HCl, 29.88 mg L-isoleucine, 56.76 mg L-leucine, 29.88 mg L-lysine-HCl, 19.92 mg L-methionine, 49.8 mg L-phenylalanine, 373.5 mg L-serine, 199.2 mg L-threonine, 39.84 mg L-tryptophan, 30 mg L-tyrosine, 19.92 mg uracil, and 149.4 mg L-valine. Store at 4°C.

3-AT Medium: for 2 M 3-AT mix 1.68 g 3-AT (Sigma) in 10 ml of H_2O, filter-sterilize and store at 4°C. For a final concentration of 5mM 3-AT add 2.5 ml stock solution to 1l appropriately supplemented SD medium after autoclaving and cooling of medium to ~50°C. Note that 3-AT medium should never contain histidine.

General remarks: For preparation of plates, add 20 g of agar per litre of medium. To reduce red pigmentation of yeast colonies due to the *ade*2 mutation, supplement the medium with higher concentrations of adenine. Alternatively, to enhance the pigmentation and visual selection through the *ade*2 mutation, reduce the concentration of adenine in the medium.

Media for *E.coli* are prepared as described by Miller (1972) and Sambrook *et al.* (1989). Both *E.coli* and yeast cells can be stored at −70°C in media containing 15–50% (v:v) glycerol.

4.9.2 *Supplies and solutions*

Acid-washed glass beads: submerge glass beads (0.5 mm diameter, Sigma G9268) in concentrated nitric acid for 1–2 h. Discard the acid and wash the beads extensively under running water with occasional stirring until the pH is near neutral as measured using pH paper. Rinse with distilled water, drain, and spread the glass beads on a tray before drying in an oven.

Carrier DNA: Prepare 10 mg/ml salmon or herring sperm DNA in TE buffer pH 8.0. Dissolve the DNA in TE buffer by drawing up and down through a 10 ml pipette; incubate and stir overnight at 4°C. Sonicate the DNA twice for 30 s with a large probe sonicator at three-quarters power. The resulting DNA should have an average size of 7 kb as judged by agarose gel electrophoresis, and range in size from 2–15 kb. Smaller, average size (1–2 kb) owing to oversonication, dramatically reduces transformation efficiency. Alternatively, DNA can be sheared by drawing it through a small diameter syringe needle 2 to 3 times. Boil carrier DNA for 20 min, cool on ice, aliquot and store at −20°C.

Cell lysis solution: 0.2 M NaOH; 1% v:w SDS. Mix sterile stock solutions of each just prior to use.

Laemmli buffer: 10% glycerol, 5% βME, 3% SDS, 1× stacking gel buffer.

Li acetate solution: 1× each of stock Li Acetate and stock TE ph 7.5. Mix just prior to use.

Li acetate 10×: 1.0 M Li acetate, adjust pH to 7.5 with dilute acetic acid. Filter-sterilize.

Neutralization solution: 3 M K acetate. Adjust to pH 4.8 with glacial acetic acid and sterilize.

PEG solution: 40% PEG 3350, 0.1 M Li acetate pH 7.5, 1× TE buffer pH 7.5. Prepare a 50% PEG stock solution in H_2O, filter-sterilize and store at room temperature. This solution is very viscous and time consuming to prepare. For convenience, prepare large volumes of approximately 500 ml. Just prior to use, mix 50% PEG stock with 10× stocks of Li acetate solution and TE buffer.

Stacking gel buffer 8×: 0.4% SDS, 0.5 M Tris-Cl pH 6.8.

TE buffers. 10×: 0.1 M Tris-Cl, 0.01 M EDTA; adjust pH to 7.5 or 8.0 with HCl. Filter-sterilize.

Z-buffer: 60 mM Na_2HPO_4, 40 mM NaH_2PO_4, 10 mM KCl, 1 mM $MgSO_4$. Prepare without mercaptoethanol or Xgal. Add 2.7 ml/litre mercaptoethanol and 1 mg/litre Xgal (from a 100 mg/ml stock in DMF) just before use.

Zymolyase solution: 2 mg/ml zymolyase 20T, 100 mM K_2HPO_4, 100 mM KH_2PO_4, 1.2 M sorbitol. Prepare zymolyase solution just prior to use, from sterile stock solutions of 1M K_2HPO_4, 1 M KH_2PO_4 and 2 M sorbitol.

References

BARTEL, P., CHIEN, C.T., STERNGLANZ, R. and FIELDS, S., 1993a, Elimination of false positives that arise in using the two-hybrid system, *Biotechniques*, **14**, 920–924.

1993b, Using the two-hybrid system to detect protein-protein interactions, in Hartley, A. D. (Ed.) *Cellular Interactions and Development. A Practical Approach*, pp. 153–179, New York: IRL Press.

BEKKAOUI, F., SCHORR, P. and CROSBY, W.L., 1993, Acetolactate synthase from *Brassica napus*: immunological characterization and quaternary structure of the native enzyme, *Plant Physiology*, **88**, 475–484.

BENNETZEN, J.L. and HALL, B.D., 1982, The primary structure of the *Saccharomyces cerevisiae* gene for alcohol dehydrogenase I, *Journal of Biological Chemistry*, **257**, 3018–3025.

BURNETTE, W.N., 1981, 'Western blotting': electrophoretic transfer of proteins from sodium dodecyl sulfate-polyacrylamide gels to unmodified nitrocellulose and radiographic detection with antibody and radio-iodinated protein A, *Analytical Biochemistry*, **112**, 195–203.

CAREY, M., KAKIDANI, H., LEATHERWOOD, J., MOSTASHARI, F. and PTASHNE, M., 1989, An amino-terminal fragment of GAL4 binds DNA as a dimer, *Journal of Molecular Biology*, **209**, 423–432.

CHEVRAY, P.M. and NATHANS, D., 1992, Protein interaction cloning in yeast – identification of mammalian proteins that react with the leucine zipper of Jun, *Proceedings of the National Academy of Sciences (USA)*, **89**, 5789–5793.

CHIEN, C.-T., BARTEL, P., STERNGLANZ, R. and FIELDS, S., 1991, The two-hybrid system: a method to identify and clone genes for proteins that interact with a protein of interest. *Proceedings of the National Academy of Sciences (USA)*, **88**, 9578–9582.

FEILOTTER, H.E., HANNON, G.J., RUDDELL, C.J. and BEACH, D., 1994, Construction of an improved host strain for two hybrid screening, *Nucleic Acids Research*, **22**, 1502–1503.

FIELDS, S. and SONG, O., 1989, A novel genetic system to detect protein–protein interactions, *Nature*, **340**, 245–246.

FIELDS, S. and SONG, O.-K., 1994, System to detect protein–protein interactions, (Abstract), No. 5,283,173, US Patent Register.

FIELDS, S. and STERNGLANZ, R., 1994, The two-hybrid system: an assay for protein–protein interactions, *Trends in Genetics*, **10**, 286–292.

FRITZ, C.C. and GREEN, M.R., 1992, Fishing for partners, *Current Biology*, **2**, 403–405.

GIETZ, R.D., STJEAN, A., WOODS, R.A. and SCHIESTL, R.H., 1992, Improved method for high efficiency transformation of intact yeast cells, *Nucleic Acids Research*, **20**, 1425.

KEEGAN, L., GILL, G. and PTASHNE, M., 1986, Separation of DNA binding from the transcription-activating function of a eukaryotic regulatory protein, *Science*, **231**, 699–704.

MILLER, J.H., 1972, *Experiments in Molecular Genetics*, Cold Spring Harbor: Cold Spring Harbor Laboratory Press.

PANG, P.P., PRUITT, R.E. and MEYEROWITZ, E.M., 1988, Molecular cloning, genomic organization, expression and evolution of 12s seed storage protein genes of *Arabidopsis thaliana*, *Plant Molecular Biology*, **11**, 805–820.

PHIZICKY, E.M. and FIELDS, S., 1995, Protein–protein interactions: methods for detection and analysis, *Microbiology Reviews*, **59**, 94–123.

SAMBROOK, J., FRITSCH, E.F. and MANIATIS, T., 1989, *Molecular Cloning: A Laboratory Manual*, Cold Spring Harbor: Cold Spring Harbor Laboratory Press.

SHERMAN, F., FINK, G.R. and HICKS, J.B., 1983, *Methods in Yeast Genetics*, Cold Spring Harbor: Cold Spring Harbor Laboratory Press.

VIDAL, M., BRACHMANN, R.K., FATTAEY, A., HARLOW, E. and BOEKE, J.D., 1996a, Reverse two-hybrid and one-hybrid systems to detect dissociation of protein-protein and DNA-protein interactions, *Proceedings of the National Academy of Sciences (USA)*, **93**, 10315–10320.

VIDAL, M., BRAUN, P., CHEN, E., BOEKE, J.D. and HARLOW, E., 1996b, Genetic characterization of a mammalian protein-protein interaction domain by using a yeast reverse two-hybrid system, *Proceedings of the National Academy of Sciences (USA)*, **93**, 10321–10326.

WANG, H.W., KOHALMI, S.E. and CUTLER, W.L., 1996, An improved method for PCR using whole yeast cells, *Analytical Biochemistry*, **237**, 145–146.

T-DNA Mediated Gene Tagging in *Arabidopsis*

JENNIFER F. TOPPING AND KEITH LINDSEY

5.1 Introduction

One of the major questions in biology is concerned with how individual genes within an organism are co-ordinately expressed to elicit the complex morphological and developmental processes which occur during the life cycle. In order to answer this central question, the genes which play determinative roles in the regulation of these processes must be isolated and characterized. The transcripts and proteins encoded by these genes are possibly present at relatively low abundances and may be limited to specific cell types at precise stages of development. This, therefore, makes the cloning of these genes by conventional methods (e.g. differential cDNA library screening or expression library screening) problematic. At present there are two strategies which are proving to be extremely useful in this area of developmental biology. The first is mutant analysis and the second is gene tagging.

In this chapter we will describe in detail the T-DNA tagging approach that we have adopted in our laboratory as a means to identify and isolate genes which are expressed differentially during plant development. We will describe the T-DNA-based vector which we have constructed and discuss its merits as both an insertional mutagen and a promoter trap. Several genes have been isolated in other laboratories using T-DNA insertional mutagenesis and we will review the progress in this area to date. Over recent years *Arabidopsis thaliana* has been adopted as one of the major dicot species to be used as a model for studies in plant gene expression and development, and we will also discuss its relative merits. The success of the T-DNA tagging strategy is dependent on an efficient means of generating transformants and we will therefore describe in detail a rapid and highly efficient method of *Agrobacterium*-mediated transformation of *Arabidopsis* which we have developed in our laboratory. Procedures for the screening of transgenic populations for putative mutants and reporter gene fusion expression will then be described. The promoter trap vector which we employ contains a promoterless *gus*A gene, the activity of which can be assayed both fluorimetrically and histochemically, and detailed protocols will be presented for both procedures. Methods for determining the number of loci and precise copy number of the T-DNA vector within an individual line will be described. Finally, we will describe a method for the cloning of tagged genes using inverse PCR (IPCR).

5.2. Gene tagging and the generation of mutants

In plants, genetic mutations are induced experimentally using either of three methods: (i) chemical treatment (e.g. soaking seeds in ethyl methanesulphonate, EMS); (ii) irradiation with ionizing radiation (e.g. γ-irradiation of seeds); and (iii) ectopic insertion of foreign DNA directly into the plant genome (insertional mutagenesis). Chemical and irradiation treatments have the advantage that they are rapid to perform and large numbers of mutants can be produced very quickly. However, if the aim is to clone the gene which is responsible for the mutant phenotype then there are two main disadvantages to this approach. First, in both treatments multiple genetic lesions are produced, making it difficult to pinpoint the mutation which is responsible for the observed phenotype. Second, isolation of the mutated gene requires the use of map-based cloning techniques: this involves RFLP analysis after out-crossing to marker lines, and in many cases the lengthy procedure of chromosome walking. It is worth noting, however, that mutants generated by chemical mutagenesis or irradiation are very useful in the identification of additional mutant alleles of a characterized phenotype, generated by insertional mutagenesis.

Insertional mutagenesis is a particularly useful technique for the isolation and characterization of the genes of function defined by mutagenesis. There are two classes of DNA sequences which are used most commonly as insertional mutagens: (i) T-DNA from *Agrobacterium tumefaciens* and (ii) transposons. If the inserted DNA results in the disruption and hence loss of function of a gene, that gene is physically tagged by the insertion element. It is then a relatively straightforward procedure to isolate the tagged gene by either library screening, plasmid rescue (if the transforming DNA has a bacterial origin of replication), or inverse PCR (IPCR). The gene tagging approach can be refined further to allow the cloning of genes which are expressed differentially during development but, which although disrupted, do not necessarily result in a mutant phenotype, through the use of promoter trapping.

5.3 T-DNA vectors as insertional mutagens

In this chapter we will focus exclusively on the methodology of gene tagging using T-DNA vectors and the reader is referred to recent reviews describing the transposon tagging system (Walbot, 1992; Topping and Lindsey, 1995).

Two broad categories of T-DNA constructs have been engineered and used as insertional mutagens and gene tags. The first category contains the left and right border regions of the T-DNA and a selectable marker gene, e.g. the *npt*II gene from Tn5, which confers resistance to the antibiotic kanamycin. The second class are those constructs in which the promoter of the selectable marker gene has been removed or a promoterless screenable marker is included adjacent to either the right or left border regions of the T-DNA. In both cases the T-DNA constructs can act as mutagens, and simultaneously as gene tags for subsequent gene isolation. The second category of constructs allows in addition, the *in vivo* detection of the tagged gene promoter activity by virtue of the creation of either transcriptional or translational gene fusions between the promoterless marker gene and endogenous plant gene promoter sequences flanking the T-DNA insert. This second category comprises the promoter trap vectors (reviewed by Walden *et al.*, 1991; Topping and Lindsey, 1995). Promoter trap vectors are invaluable for the analysis of gene expression in mutants in which the insertion of the vector results in loss-of-function mutations that are lethal in the homozygous state, because gene expression can be analysed in viable heterozygotes. There is a third class of vectors which is a modification of the promoter trap vectors, used to create so-called 'gain-of-function mutants'. There are two groups within this class and collectively they are known as constructs for 'activation tagging'. In the first group the promoterless reporter gene is replaced with a promoterless growth regulator gene; here *in vivo* gene fusion results in the ectopic expression of the growth regulator gene thereby producing transformants with aberrant phenotypes (Hewelt *et al.*, 1994). The second approach to activation tagging is to introduce a strong promoter sequence randomly into the plant genome resulting in the overexpression of endogenous plant genes. Using this technique auxin and cytokinin autotrophs, and mutants displaying altered polyamine metabolism were identified (Hayashi *et al.*, 1992). For further details on this relatively new approach the reader is referred to a recent review by Walden *et al.* (1994).

In our laboratory we have constructed the promoter trap vector pΔgusBin19. This vector is based on the binary vector Bin19 (Bevan, 1984) and comprises a selectable *npt*II gene near the T-DNA right border, under the transcriptional control of the *nos* promoter and polyadenylation sequences; while at the left border is a promoterless *uid*A (*gus*A) coding region (encoding β-glucuronidase; Jefferson *et al.*, 1987) plus *nos* polyadenylation sequence (Figure 5.1). The 5′ end of the *gus*A gene is adjacent to the left border and has its own translation initiation codon. Since there are translational stop codons in the left border region in all reading frames, gene fusions generated upon integration are expected to be primarily transcriptional rather than translational, which has been confirmed by our experimental data (unpublished data). This vector has proved to function very successfully as a gene tag and promoter trap in *Arabidopsis*, tobacco and potato plants (Topping *et al.*, 1991; Lindsey *et al.*, 1993; Topping *et al.*, 1994, Rooke 1995).

There are a number of different vectors which have been constructed in other laboratories as promoter traps and they are listed in Table 5.1. The detailed protocols presented below are those which are in use in our laboratory where we have

Table 5.1 T-DNA vectors used in gene tagging

Name	Right or left border gene fusion	Selectable marker	Screenable marker	Reference
pΔgusBin19	left	*npt*II	Δ[1]*gus*A	Topping *et al.* (1991)
pPCV621	right	Δ*aph*(3′)II	none	Koncz *et al.* (1989)
pGV1047	right	*npt*II	Δ*gus*A	Kerbundit *et al.* (1991)
pPRF120	right	*npt*II	Δ*gus*A	Fobert *et al.* (1991)
pMOG553	right	*hpt*	Δ*gus*A	Goddijn *et al.* (1993)
pMHA2	right	*npt*II	Δ*gus*A	Mandal *et al.* (1995)

[1]Δ = promoterless gene.

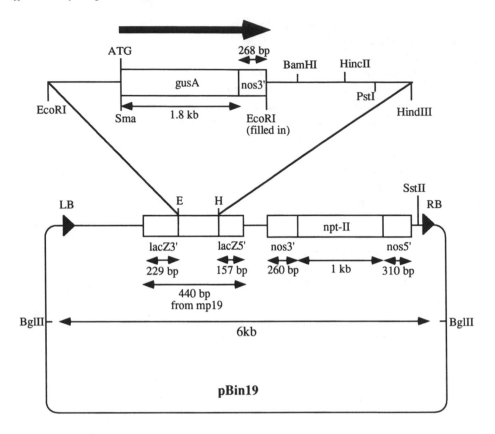

pΔgusBin19

12 kb

Figure 5.1   The *gusA* coding region plus the *nos* termination sequence is inserted into the multiple cloning region of pBin19 (Bevan, 1984). *LB* = T-DNA left border repeat; *RB* = T-DNA right border repeat. The restriction enzyme sites indicated are unique within the T-DNA insert and are therefore useful for the generation of border fragments. The sizes (in base pairs) of each of the genes and regulatory sequences are shown. The *gusA* ATG codon is 673 bp from the T-DNA *LB*.

worked exclusively with pΔgusBin19. However, in principle the protocols could be used with other T-DNA based vectors. Where differences may arise when working with other vectors, this will be pointed out. Table 5.2 lists genes which have been cloned to date using T-DNA tagging vectors.

5.4   T-DNA promoter trap vectors

The potential advantages of using promoter trap vectors as tools with which to tag and clone differentially expressed genes in plants are becoming widely appreciated in many different areas of plant research, and over recent years these potential advantages have begun to be realized. After a population of promoter trap lines has been

Table 5.2. Genes isolated from *Arabidopsis* by T-DNA tagging

Gene	Function	Reference
Agamous	homoeotic gene	Yanofsky *et al.* (1990)
Glabrous GL1 + GL2	trichome development	Marks and Feldmann (1989); Herman and Marks (1989); Larkin *et al.* (1993)
PFL	leaf development	van Lijsebettens *et al.* (1991)
FL1	flower homoeotic gene	den Boer *et al.* (1991)
LD	control of flowering time	Lee *et al.* (1994)
FEY	oxidoreductase	Callos *et al.* (1994)
DIM	cell elongaton	Takahashi *et al.* (1995)
dwf	cell elongation	Feldmann *et al.* (1989)
FAD2	fatty acid desaturation	Okuley *et al.* (1994)
FUS6	signal transduction	Castle and Meinke (1994)
FUS1/COP1	signal transduction	Deng *et al.* (1992)
FUS7/COP9	signal transduction	Wei and Deng (1993)
HY4	DNA photolyase	Ahmad and Cashmore (1993)
TSL	protein kinase	Roe *et al.* (1993)
SPY	GA signal transduction	Jacobsen and Olszewski (1993)
HVT1	nucleic acid helicase	Wei *et al.* (1997)

generated via *Agrobacterium*-mediated transformation it is then a relatively simple, albeit time consuming, procedure to screen populations for individual lines exhibiting specific temporal or spatial *in vivo* activation of the reporter gene (usually a promoterless *gus*A gene).

We will describe briefly two areas of plant research which have benefited from adopting this approach; first, plant development, specifically embryogenesis and gene expression within vascular tissues; and second, the response of plants to stress, either pathogen attack or abiotic stress.

In our laboratory, where we have generated over 2100 independent transgenic *Arabidopsis* lines containing pΔgusBin19, two screening programmes were initiated. The first was to identify transformants which exhibited GUS activity in the embryos; and the second was to identify transformants in which GUS activity was localized to the vascular tissues. Both screens were relatively straightforward to carry out and both fluorimetric and histochemical GUS assays were used (described in detail below). In both cases a number of lines were characterized which exhibited interesting patterns of GUS expression; the T-DNA tagged genes from several of these lines have subsequently been cloned, and the wild-type alleles isolated (Muskett *et al.*, 1994; Topping *et al.*, 1994; Wei *et al.* 1997).

Differential gene expression in response to nematode attack in *Arabidopsis* has also been studied using T-DNA tagged lines. Individual transformants were infected with the pathogen and then screened for induced GUS activity in the feeding structures within the infected roots (Goddijn *et al.*, 1993). T-DNA tagged *Arabidopsis* lines have also been identified which exhibit elevated levels of GUS activity when the plants are cold shocked or treated with abscisic acid (Mandal *et al.*, 1995).

5.5 *Arabidopsis* as a model plant species

The majority of developmental studies in plants have focused on a small number of so-called model species. Maize (*Zea mays*) is the preferred monocot, and tobacco (*Nicotania tabacum*) and *Arabidopsis thaliana* are the major dicot model species. However, over the past few years, an increasing amount of research into plant developmental processes has been carried out in *Arabidopsis* (Meyerowitz, 1987). This is due to *Arabidopsis* having a number of key features which make it an ideal species for the isolation of low-abundance regulatory genes via gene tagging and mutant analysis:

1.   A small genome (haploid size 70 000–100 000 kb, five times the size of that of the yeast *Saccharomyces cerevisiae* and containing low abundance of repetitive DNA.

2.   A well defined and rapid (about 6 week) life cycle.

3.   The flowers self-pollinate and each plant will produce approximately 10 000 seeds when grown under favourable conditions.

4.   Amenability to *Agrobacterium*-mediated transformation.

5.   The plants are relatively small and can be grown to maturity in 4 × 4cm pots in natural or artifical light, or *in vitro*.

All these features make *Arabidopsis* a good candidate for large mutagenesis screens. However, *Arabidopsis* does have some limitations. Due to the small size of the flowers, carrying out crosses manually is more of an art than a science; and if your interests lie in an aspect of plant development which does not occur in *Arabidopsis*, e.g. tuberization, this species will clearly be of little use. However, in the majority of cases *Arabidopsis* can be used as a model, and we will focus exclusively on T-DNA mediated gene tagging in *Arabidopsis* in this chapter.

5.6 Protocols

PROTOCOL 5.1   THE INTRODUCTION OF T-DNA VECTORS INTO *ARABIDOPSIS THALIANA* VIA *AGROBACTERIUM*-MEDIATED TRANSFORMATION

Direct gene transfer systems have been developed for *Arabidopsis* (Damm and Willmitzer, 1991) but here we will only consider transformation via infection with *Agrobacterium tumefaciens*. Transformation protocols have been developed for whole-plants and excised tissues of *Arabidopsis*. To date, there are three whole plant (non-tissue culture) systems: first, the intact seed method (Feldmann and Marks, 1987), which has been used to generate approximately 13 000 independent transformants to date (Forsthoefel *et al.*, 1992), but the technique is not easily reproduced in other laboratories; second, meristem inoculation (Chang *et al.*, 1994); and third, whole-plant vacuum infiltration (Bechtold *et al.*, 1993). The plant tissues most commonly transformed are cotyledons (Schmidt and Willmitzer, 1991) and roots (Valvekens *et al.*, 1992). We (Clarke *et al.*, 1992) have modified the root transformation system of Valvekens *et al.* such that, in our hands, the frequency of recovery of transformants is increased up to 100-fold; we have generated approximately 2100 independent transformants using this method, and it has been used successfully in other laboratories.

Materials

1. Bleach solution: 5% (v/v) commercial bleach containing 0.05% (v/v) Tween 20.

2. 70% (v/v) ethanol.

3. Silver thiosulphate stock solution (1.25 mg/ml): add dropwise a solution of silver nitrate (2.5 mg/ml) to an equal volume of a solution of sodium thiosulphate pentahydrate (14.6 mg/ml). Filter sterilize.

4. Germination medium (GM): half strength Murashige and Skoog medium (Murashige and Skoog, 1962; Sigma M5519), 10 g/litre sucrose. Adjust pH to 5.8 with 1 M KOH, add 8 g/litre Difco Bacto agar, autoclave (121°C, 15 min) to sterilize, add 5 mg/litre silver thiosulphate solution (filter sterilized) before plating out.

5. Stock solution of 4 mg/ml 2,4-dichlorophenoxyacetic acid (2,4-D) in dimethylsulphoxide (DMSO).

6. Stock solution of 0.4 mg/ml kinetin in DMSO.

7. Callus-inducing medium (CIM): Gamborg's B5 medium (Gamborg *et al.*, 1968; Sigma), 0.5 g/litre MES, 20 g/litre glucose. Adjust pH to 5.8 with 1 M KOH, add 8 g/litre Difco Bacto agar, autoclave (121°C, 15 min) to sterilize. After autoclaving, add 5 mg/litre silver thiosulphate solution (filter sterilized), 0.5 mg/litre 2,4-D, 0.05 mg/litre kinetin.

8. Stock solution of 200 mg/ml vancomycin (filter sterilized).

9. Stock solution of 50 mg/ml kanamycin sulphate (filter sterilized).

10. Stock solution of 20 mg/ml 2-isopentenyladenine (2-iP) in DMSO.

11. Stock solution of 1.2 mg/ml indole-3-acetic acid (IAA) in DMSO.

12. Shoot-inducing medium (SIM): Gamborg's B5 medium (Sigma), 0.5 g/litre MES, 20 g/litre glucose. Adjust pH to 5.8 with 1 M KOH, add 8 g/litre Difco Bacto agar, autoclave (121°C, 15 min). After autoclaving, add 850 mg/litre vancomycin, 35 mg/litre kanamycin sulphate, 5 mg/litre 2-iP, 0.15 mg/litre IAA.

13. Shoot overlay medium (SOM): as SIM, but replacing the agar with 8 g/litre low melting point agarose (SeaPlaque, FMC); autoclave (121°C, 15 min).

14. Shoot elongation medium (SEM): half strength Murashige and Skoog Medium (Sigma), 10 g/litre sucrose. Adjust pH to 5.8 with 1 M KOH, add 8 g/litre Difco Bacto agar, autoclave (121°C, 15 min).

15. Compost: From Russell *et al.* (1991): Irish moss peat (Joseph Bently & Sons, Barrow-on-Humber, UK), John Innes potting compost No. 3, horticultural potting grit (Joseph Bently & Sons, Barrow-on-Humber, UK), coarse vermiculite (Vermiperl Medium Grade, Silverperl Ltd, Gainsborough, UK). Mix in a ratio of 2:2:2:1, autoclave (121°C, 30 min).

16. LB (Luria–Bertani) medium: 10 g/litre bacto-tryptone, 5 g/litre bacto-yeast extract, 10 g/litre NaCl. Adjust pH to 7.5 with 1 M NaOH, autoclave at 121°C for 15 min.

17. Filter-sterilized stock of rifampicin: 20 mg/ml in methanol.

18. *Agrobacterium* culture medium: sterile LB (Luria–Bertani) medium, 50 mg/litre kanamycin sulphate (or other appropriate selective agent), 100 mg/litre rifampicin (or other appropriate selective agent).

19. *Agrobacterium* culture dilution medium: Gamborg's B5 salts solution (Sigma), 20 g/litre glucose, 0.5 g/litre MES. Adjust pH to 5.7 with 1 M KOH, autoclave at 121°C for 15 min.

20. Sterile double-distilled water (autoclave 121°C, 15 min).

21. Micropore® gas-permeable tape (3M).

22. 9-cm Petri dishes (Falcon 3003).

23. Nylon or stainless steel mesh (100 μm pore diameter).

24. Sterile filter paper.

25. Polypot containers (SC020, Northern Media Supply Ltd, UK).

26. Gas-permeable transparent SunCap® film (C6920, Sigma).

27. Sterile perlite (autoclave 121°C, 15 min).

28. Sterile Gamborg's B5 salt solution (autoclave 121°C, 15 min).

29. Aracon® tubes (Beta Developments, Gent).

30. Sterile transfer pipettes.

Method

The *Arabidopsis* roots to be transformed are harvested from *in vitro* light-grown seedlings.

Seed sterilization and seedling growth

1. *Arabidopsis* seeds should be stored at 4°C for 1 week before sterilization, to maximize germination frequency.

2. Surface sterilize the seeds by:

    (a) immersion in 70% (v/v) ethanol for 30 s; then remove the ethanol with a transfer pipette; then

    (b) replace with 5% (v/v) commercial bleach solution containing 0.05% Tween 20 as a surfactant and leave for 15–20 min.

3. Remove bleach solution with a transfer pipette and wash the seeds four times in sterile double-distilled water.

4. Suspend the seeds in a drop of sterile water and transfer them to the surface of GM agar plates, using a transfer pipette. Spread out the seeds evenly (approximately 40 per 9-cm Petri dish) and seal the plates with Micropore® tape.

5. Leave the plates for 7 days at 4°C to maximize germination.

6. Incubate the plates for 3–4 weeks in a standard light regime for cultures (at least 25–50 μmol of photons per m^2/s) to allow root growth.

Root explantation and inoculation with Agrobacterium

1. After 3 weeks, when the seedlings have produced abundant (but not yet greening) roots (see section 5.7, Note 1), gently remove them from the agar and excise the root systems.

2. Place the intact root systems on the surface of agar plates containing CIM, and incubate at approximately 20–22°C for 3 days.

3. Remove the roots from the CIM and place in a sterile Petri dish. Cut the roots into segments of approximately 0.5 cm in length.

4. Grow up a liquid suspension of *Agrobacterium* cells at 29°C on a rotary shaker (200 rpm) for 48 h, then dilute with sterile *Agrobacterium* dilution medium, to a final optical density of 0.1 (600 nm).

5. Place root explants in 20 ml *Agrobacterium* culture medium and incubate at room temperature for 2 min. Remove the roots onto a nylon or stainless steel mesh (100 μm

pore diameter) and drain off the bacterial suspension, then remove excess liquid by blotting the roots on sterile filter paper.

6. Replace the roots on the CIM agar plates, and culture for 48–72 h.

7. Excess bacterial growth should then be removed by washing the root segments with sterile *Agrobacterium* culture dilution medium (see section 5.7, Note 2), using a nylon/steel mesh, followed by blotting the roots dry on sterile filter paper.

Selection and regeneration of transformants

1. Melt some SOM and leave it to cool to about 30°C. Suspend the roots of approximately 40 seedlings (approximately 0.1 g tissue, corresponding to about 300 individual root segments) per 10 ml molten SOM.

2. Pour 10 ml of the root/SOM suspension into plates (9 cm) containing 40 ml SIM. Ensure that the root explants are evenly dispersed. Seal the plates with Micropore® tape (see section 5.7, Note 3), and incubate at approximately 25°C in fluorescent light of at least 25–50 µmol/m²/s.

3. After approximately 3 weeks, the first putative transformants begin to appear on the roots, which will appear as green calluses. From these growths, kanamycin-resistant shoots will regenerate over a period of several weeks.

4. When the shoots have expanded leaves, transfer them to 60 ml polypots containing 20 ml SEM. When the plantlets bolt, prior to flowering, remove the lid of the polypot and replace it with gas-permeable transparent SunCap® film, (see section 5.7, Note 4). Under these conditions, at least 80% of the transformed shoots are expected to flower and set seed (see section 5.7, Notes 5 and 6).

Bulking up of T_2 plants

1. Add sterile Gamborg's B5 salt solution to a layer of sterile perlite in a Petri dish until the perlite is saturated, pour off any excess.

2. Plate out the seeds (approximately 20 per plate) from the T_1 plants and incubate for 7 days at 4°C. Then place them in a culture room at approximately 22°C (see section 5.7, Note 7).

3. When the seedlings are large enough to handle they can be transferred to sterile, moist compost. Cover the seedlings with cling film or place them in a micropropagator for the first week (at approximately 20°C), and gradually harden off the plants by increasing ventilation over the following week.

4. Just before the plants bolt, cover them with Aracon tubes (Beta Developments, Gent) and collect seed over the following few weeks (see section 5.7, Note 8).

PROTOCOL 5.2 SCREENING FOR T-DNA TAGGED MUTANTS

Due to the low number of T_1 seeds produced and the hemizygous state of the transgene, it is the T_2 population which is screened for mutant phenotypes. It is therefore necessary to screen more than one family from each line to allow for segregation of the mutant phenotype. The method of screening for mutant phenotypes depends on the type of mutant (developmental, biochemical, physiological) of interest. Evidence that the mutation may be due to the insertion of a T-DNA requires the demonstration of co-segregation of the T-DNA and the mutant phenotype in progeny from selfing and outcrossing with wild-type lines. The first step is to demonstrate linkage between a dominant selectable marker, e.g. kanamycin resistance conferred by the T-DNA and the mutant phenotype.

Analysis of the segregation of kanamycin resistance with a mutant phenotype

Materials

1. Bleach solution (see protocol 5.1).
2. 70% (v/v) ethanol.
3. Sterile 9-cm Petri dishes (Falcon 3003).
4. Micropore® tape.
5. Stock solution of 50 mg/ml kanamycin sulphate (filter sterilized).
6. Germination medium (GM) (see protocol 5.1(4)), but omit the silver thiosulphate. After autoclaving and cooling to approximately 50°C, add kanamycin sulphate solution to a final concentration of up to 150 mg/litre (see section 5.7, Note 9).

Method

1. Plate out approximately 200 sterilized T_2 seeds (see protocol 5.1) on GM plates with and without kanamycin selection. A T-DNA present as a single active locus will segregate 3 kan^r : 1 kan^s in the progeny. The mutant phenotype, if caused by a single recessive mutation, will appear in 25% of the seedlings on GM medium which lacks kanamycin. If the mutant is embryonic lethal, all the viable progeny will be phenotypically wild-type (heterozygous for the mutation or wild type).

2. Pick approximately six mutants, transfer them to soil and grow to maturity under greenhouse conditions. Collect the T_3 seed from the mutants (assuming they are not embryonic- or seedling-lethal or sterile) and plate out approximately 500 on kanamycin selection plates. If the mutant phenotype and T-DNA are linked, 100% of the seedlings will be kanamycin resistant (homozygous for the *npt*II gene) and 100% mutant phenotype (homozygous for the mutant gene).

3. If it is not possible to grow the mutants to maturity, then kanamycin-resistant but phenotypically wild-type T_3 seedlings should be analysed. They should segregate, on selfing, 25% kanamycin-resistant mutants (homozygous for T-DNA and mutant gene); 50% kanamycin-resistant wild-type phenotypes (heterozygous for T-DNA and mutant gene); and 25% kanamycin-sensitive wild-type phenotypes (lacking both T-DNA and mutant gene; see section 5.7, Note 10). Southern blots should be carried out to confirm this.

4. To determine whether an embryonic-lethal mutation is caused by a T-DNA, the T_3 seeds are collected from a kanamycin-resistant plant (which is expected to be heterozygous for both the mutant gene and the T-DNA) and plated out on kanamycin selection plates. Twenty-five per cent of the seeds should fail to germinate (homozygous for the embryonic lethal mutation); of the germinating seeds, 66% of the seedlings should be kanamycin-resistant wild-type phenotypes (heterozygous for T-DNA and mutant gene) and 33% should be kanamycin-sensitive wild-type phenotypes (lacking both T-DNA and mutant gene).

5. Further evidence of co-segregation can be obtained by outcrossing lines that are putatively homozygous for both T-DNA and mutant gene to wild-types. This is carried out as follows:

(a) Remove the anthers from unopened buds of the female parent (see section 5.7, Note 11).

(b) Remove all other buds close to the buds to be fertilized to avoid confusion.

(c) Pollinate with dehiscing anthers removed from the male parent.

(d) Repeat the pollination the following day (see 5.7, Note 12).

Collect F_1 seed from the cross (which is expected to be heterozygous for both the mutant gene and T-DNA). Grow up and allow the F_1 generation to self, collect the F_2 seed and plate it out on kanamycin selection plates. Twenty-five per cent of the seedlings are expected to be kanamycin-resistant mutants (homozygous for T-DNA and mutant gene); 50% kanamycin-resistant wild-type phenotypes (heterozygous for T-DNA and mutant gene); and 25% kanamycin-sensitive wild-type phenotypes (lacking both T-DNA and mutant gene). Southern blots should be carried out to confirm this.

PROTOCOL 5.3 MOLECULAR ANALYSIS OF T-DNA TAGGED MUTANTS

After confirming that the mutant phenotype co-segregates with the selectable marker on the T-DNA, the next step is to carry out detailed molecular analysis on the mutants. By probing the genomic DNA of the mutants (or heterozygotes, for example, if the mutant phenotype is seedling lethal) with fragments of the T-DNA, it is possible to determine:

1. The precise copy number and arrangement of the T-DNA(s) present in the mutant.

2. The size of border fragments (DNA fragments containing a region of the T-DNA juxta-posed to genomic DNA) which may be suitable for cloning (see Protocol 5.6).

Genomic DNA is prepared using standard techniques (e.g. Lindsey and Jones, 1989), digested with the appropriate restriction enzyme(s) (see below), size fractionated by electrophoresis, and blotted onto nylon membranes, again using standard techniques (described in Sambrook *et al.* 1989). The Southern blots are then probed with T-DNA fragments using standard hybridization techniques (e.g. Church and Gilbert, 1984)

It is possible to gain information about the number and organization of the T-DNA(s) within a particular transformant by carefully selecting the restriction enzymes with which the plant genomic DNA is cut prior to electrophoresis. There are two commonly used methods to estimate the number of T-DNA copies within a transformant. In the first, an internal fragment of the T-DNA is excised by digesting the genomic DNA with restriction enzymes that cut twice within the target sequence. Following autoradiography, the copy number is estimated by calibrating the intensity of the internal fragment band against samples of known copy numbers of the foreign DNA run on the gel. Each track of the gel must contain an equal amount of genomic DNA for accurate quantification, and genomic DNA from untransformed plants must be added to the copy number control samples, to allow for any non-specific background hybridization. In the second method, border fragments are generated by digesting the genomic DNA with a restriction enzyme known to cut at only one site within the target DNA; the second site of cleavage is in the genomic DNA rather than the target DNA. Since the site of the second cleavage is expected to be different for each independent insertion event, the number of bands observed is a good indication of the number of copies present. Similarly, tandem arrays of a gene construct can be diagnosed by cutting with single enzymes that cut once within the construct. In this case the expected result is a relatively intense band, representing artificially generated internal fragments within the tandem repeat, each of the same size, plus the border fragments, of different sizes. This experiment is important in order to differentiate between single and multiple insertion events when segregation data have indicated a single locus insertion. It is always important to run a negative control of non-transformed genomic DNA, to check that, under the experimental conditions used, there is no background hybridization which could confuse the analysis.

Screening for GUS fusion activity

As discussed above, by using a promoterless *gusA* reporter gene as an integral part of the T-DNA, genes can be tagged and isolated in the absence of an associated phenotype. The absence of an abnormal phenotype in transformants homozygous for a T-DNA insertion can be due to a number of possible reasons. There may be some redundancy of gene function; or the expression of a phenotypic aberration may be dependent on certain environmental conditions. Depending on the developmental process being studied, GUS activity can be screened for by either histochemistry or by fluorimetry.

PROTOCOL 5.4. ASSAY OF β-GLUCURONIDASE (GUS) ACTIVITY BY FLUORIMETRY

This assay is quantitative and it can readily be carried out in conjunction with a rapid protein assay to allow accurate quantification of the data. Plant material can be frozen prior to the assay at $-80°C$ without any loss in activity (see section 5.7, Note 13).

Materials

1. GUS buffer: 50 mM sodium phosphate buffer (pH 7.5), 10 mM β-mercaptoethanol, 0.1 % (v/v) Triton X-100 and 1 mM EDTA.

2. Acid-washed sand (Sigma).

3. 5 mM 4-methylumbelliferone glucuronide, MUG (Sigma).

4. 4-methylumbelliferone, 4-MU (Sigma).

5. 0.2 M Na_2CO_3.

Method

1. In an Eppendorf tube, grind the tissue (10–100 mg) in 200 μl of GUS buffer, with a small amount of acid-washed sand. After vortexing thoroughly, leave the tube on ice until all the samples have been ground.

2. Pellet the cell debris by centrifugation in a microfuge (12 000 rpm, 5 min).

3. To a fresh Eppendorf tube which contains 0.7 ml of GUS buffer, transfer 100 μl of the supernatant. Save the remainder for a protein assay (see below).

4. Equilibrate the samples at 37°C for 5 min.

5. To each sample add 200 μl 5 mM MUG made up in GUS buffer (i.e. at a final concentration 1 mM in the reaction mix). The substrate can be made up in advance and stored in the dark at $-20°C$.

6. Incubate the samples at 37°C.

7. At various known time points (e.g. 0, 10, 20, 60 and 90 min) take 100 μl samples and add directly to 0.9 ml of 0.2 M Na_2CO_3 (see section 5.7, Note 14). The samples can be stored at room temperature in the dark until the last time points have been taken.

8. Assay the fluorescence by removing 200 µl aliquots of the stopped reactions into a 96-well microtitre plate. Measure the fluorescence at an excitation wavelength of 365 nm and an emission wavelength of 455 nm.

9. By measuring the fluorescence of serial dilutions of the reaction product, 4-MU (Sigma), a standard curve can be constructed. Thus, in conjunction with the protein estimation detailed below, the enzyme activity per sample can be expressed as pmoles of 4-MU produced/mg protein per minute.

Assay of protein content in tissue homogenate

Materials

1. Bradford's reagent (BioRad)

2. Sterile distilled water (SDW)

3. Bovine serum albumin. (BSA)

Method

1. Remove 5 µl of the crude cell extract (step 3 of the protocol above) and add directly to a 1 ml cuvette containing 200 µl of Bradford's reagent plus 795 µl of SDW; mix well.

2. Leave at room temperature for 5–10 min. The samples can be left for up to an hour but no longer because the protein precipitates in the GUS buffer.

3. Read the absorbance, against a blank, at 595 nm. BSA can be used to construct a calibration curve.

PROTOCOL 5.5 HISTOCHEMICAL LOCALIZATION OF GUS ACTIVITY

The expression pattern of *gus*A gene fusions in stably transformed plants can be analysed very precisely, both temporally and spatially, by virtue of a simple histochemical assay. The GUS activity can be localized accurately *in vivo* using X-Gluc (e.g. from Biosynth) as a substrate. The reaction takes place in two steps: first the glucuronic acid moiety is cleaved to generate a colourless indoxyl intermediate, which is in turn oxidized to an insoluble blue precipitate. There are three main advantages in using GUS for this type of experiment: (i) a low background GUS activity in the majority of plant tissues (but see Wilkinson *et al.*, 1994); (ii) under appropriate experimental conditions the product accumulates within the plant cells where the gene is expressed and is non-diffusible; (iii) the substrate is readily taken up into the plant cells by simple diffusion or following vacuum infiltration of tissues.

There are, however, some disadvantages: (i) the assay is not readily suitable for use on living tissues, although some progress is being made in the development of non-lethal assay techniques (Kirchner *et al.*, 1993); (ii) an unrealistic picture can be given of the steady-state abundance of the native protein for which GUS is being used as a reporter, because the enzyme and product may be more stable in some tissues; (iii) in tissues where there is a high level of GUS expression, some leakage of the colourless reaction intermediate has been shown to occur; however, this can be minimized by the addition of either or both the oxidative agents potassium ferri- and ferrocyanide to a final concentration of up to 5 mM in the substrate solution (Stomp, 1990; Mascarenhas and Hamilton, 1992); and (iv) as with all histochemistry, the results are not quantitative.

Depending upon the individual tissues and organs involved, the preparation of the plant material for staining will vary. For example, the roots, flowers and leaves of *Arabidopsis* and the roots of tobacco seedlings can be stained without any prior treatment. However, stems and leaves of, for example, tobacco and potato plants should be cut into thin (1–3 mm) sections prior to staining. When working with larger tissue samples, vacuum infiltration is sometimes useful.

Materials

1. X-Gluc: 20 mM stock in *N-N*-dimethylformamide, stored in 100 μl aliquots at −20°C.

2. Substrate solution: 1 mM X-Gluc in 100 mM sodium phosphate buffer (pH 7.0) containing 10 mM EDTA, 1–5 mM each of potassium ferri- and ferrocyanide, and 0.1% Triton X-100.

3. 95% (v/v) ethanol.

Method

1. Place the tissue in an Eppendorf tube or a well of a microtitre plate and cover with 100–200 μl of substrate solution.

2. If the incubation is being carried out in a microtitre plate, ensure that the sample is well sealed to prevent the samples drying out.

3. Incubate at 37°C overnight (or for shorter periods if appropriate).

4. If necessary, soak the tissue in 95% ethanol to clear the tissue of chlorophyll; the blue precipitate is stable in ethanol.

5. If the tissue is to be fixed for sectioning, begin the fixation process after the overnight incubation.

Isolation of T-DNA tagged genes

Once a T-DNA tagged mutant or promoter trap line of interest has been identified, the next step is to clone the genomic region flanking the T-DNA sequence. This DNA fragment can subsequently be used as a probe to screen a genomic library in order to isolate the wild-type allele. The best method to isolate the tagged gene will depend on the T-DNA vector which has been used; for example, some vectors contain a bacterial origin of replication for plasmid rescue of tagged sequences (Koncz *et al.*, 1989). Here, we will describe a method based on the inverse polymerase chain reaction (IPCR; Ochman *et al.*, 1988; Innis *et al.*,1990), which can be used to clone the flanking genomic DNA no matter which T-DNA construct has been used.

It is extremely advantageous if the tagged line contains only a single T-DNA insert. Therefore, if multiple inserts are present, it is well worth outcrossing the tagged line with wild-type lines in an attempt to segregrate out the T-DNAs. This, however, will not be successful if two T-DNAs are closely linked and means that the isolation of the tagged gene will be more tedious, although not impossible (see section 5.7, Note 15). The IPCR method is very simple and rapid to perform if a single T-DNA is present. For this method to work successfully, it is necessary to generate a border fragment of a suitable size by restriction enzyme digestion (see below). If this is not possible, a genomic library can be made and screened using a fragment of the T-DNA as a probe. Alternatively, 5′ RACE PCR (Frohman, 1990) can be used to amplify that region of the fusion transcript, generated *in vivo*, between the promoterless reporter and native gene mRNA, using primers based on the reporter gene sequence. In our laboratory we have recently used this approach successfully. We used the 5′ RACE System (Gibco BRL) and followed the manufacturer's instructions.

PROTOCOL 5.6   INVERSE PCR AMPLIFICATION OF T-DNA FLANKING SEQUENCES

To amplify successfully T-DNA flanking sequences by IPCR, it is essential to identify restriction fragments containing T-DNA sequences and plant flanking DNA, such that the expected PCR product is no greater than approximately 2.5–3 kb. Larger products are difficult to amplify. Purified genomic DNA from transgenic plants is analysed by Southern hybridization following digestion with a range of restriction enzymes designed to generate appropriate T-DNA/plant DNA border fragments (see Protocol 5.3). It is best to use restriction enzymes which cut once within the T-DNA at a known location. Monomeric circles are then generated by religating the fragments under specific conditions. Primers can be designed to amplify the genomic DNA at either the left or the right border of the T-DNA. In promoter trap lines the region upstream from the left border is likely to include the promoter region of the tagged gene, whereas the region downstream from the right border should include the coding region of the gene (but see section 5.7, Note 16). Below are listed the primers which we designed in our laboratory to amplify both 3′ and 5′ regions flanking pΔgusBin19; the relative position of the primers are illustrated in Figure 5.2 (see section 5.7, Note 17). The entire T-DNA of pΔgusBin19 has been completely sequenced (Wei *et al.*, 1994). Obviously different primers may have to be designed if another vector is used.

Materials

1. 5× ligation buffer: 250 mM Tris-Cl pH 7.4, 50 mM MgCl$_2$, 50 mM dithiothreitol, 5 mM ATP.

2. T4 ligase: 1 unit/ml (BRL/Gibco, Middlesex, UK).

3. 10× PCR buffer (contains 1.5 mM Mg^{2+}; Promega, Southampton, UK).

4. 2 mM dNTP.

5. Primers (100 ng/ml). The following primers are for use with the vector pΔ*gus*Bin19:

    LB[-267] 5′ CAACACTCAACCCTATCTCGGGC 3′

    LB[-169] 5′ CCAGCGTGGACCGCTTGCTGCAAC 3′

    LB[-63] 5′ CGTCCGCAATGTGTTATTAAG 3′

PRIMERS FOR IPCR OF T-DNA FLANKING SEQUENCES

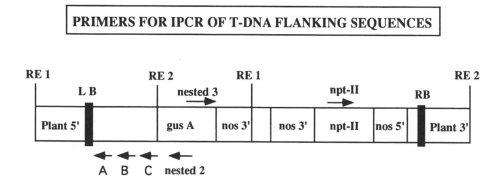

Figure 5.2   The relative positions of the oligonucleotide primers (arrows) for the amplification of T-DNA flanking sequences (Plant 5′ and Plant 3′) by IPCR are indicated schematically. RE = restriction enzyme site; LB = T-DNA left border repeat; RB= T-DNA right border repeat; A = primer *LB* [−63]; B = primer *LB* [−169]; C = primer *LB* [−267]. The relative positions of the *gus*A and *npt*II genes and the 3′ and 5′ *nos* sequences are shown.

nested 2: 5' CAGGACGTAACATAAGGG 3'

nested 3: 5' GACTGGCATGAACTTCG 3'

nptll: 5' GTCATAGCCGAATAGCCTC 3'

The relative positions of the primers are illustrated in Figure 5.2.

6.   25 mM $MgCl_2$.

7.   *Taq* polymerase: 5 units/ml (Promega, Southampton, UK).

Method

1.   Ligate digested DNA fragments (2 μg in 40 μl) in 200 μl 5× ligation buffer plus 14 μl T4 ligase, made up to a final volume of 1 ml with sterile distilled water.

2.   Extract 500 μl ligated DNA with phenol/chloroform, precipitate the DNA in ethanol and resuspend to a concentration of 20 ng/ml.

3.   50 ng DNA (2.5 μl) is used as substrate for IPCR amplification in the following reaction mixture: 2.5 ml 10× PCR buffer (Mg^{2+} free), 1 μl each primer (100 ng/ml), 2.5 ml 20 mM $MgCl_2$, 0.4 ml *Taq* polymerase (5 units/ml); made up to 22.5 ml with sterile distilled water (see section 5.7, Note 18).

4.   Pipette the reaction mixture into an Eppendorf tube, cover with 1 drop of paraffin oil and place the tube in a thermal cycler, holding the temperature at 80°C. To start the reaction, add 2.5 μl 2mM dNTP (to give a final reaction volume of 25 μl).

5.   Carry out PCR amplification over 30 cycles using the following conditions: denaturation at 95°C for 4 min during the first cycle and for 1 min in subsequent cycles; primer annealing at 55°C, 2 min; extension by *Taq* polymerase at 72°C for 10 min during the first cycle and for 3 min in subsequent cycles; denaturation at 95°C, 1 min; and final extension at 72°C, 10 min (see section 5.7, Notes 19 and 20).

6.   Separate the reaction products in a 1% (w/v) agarose thin gel and stain with ethidium bromide to visualize the DNA.

7.   The amplified products can be simply and efficiently cloned into a commerically available vector pCRII® (contained in the TA cloning® kit manufactured by Invitrogen, San Diego) which is designed specifically for this purpose.

5.7   Notes and comments

To Protocol 5.1: The introduction of T-DNA vectors into *Arabidopsis thaliana* via *Agrobacterium*-mediated transformation.

1.   Roots which are beginning to green are too old to respond to the regeneration conditions.

2.   Following inoculation with *Agrobacterium*, wash with *Agrobacterium* culture dilution medium to remove as much excess liquid as possible from the roots. This is to prevent an overgrowth of bacteria on the roots.

3.   Micropore® tape allows gas exchange, preventing the accumulation of moisture inside the Petri dishes.

4.   SunCap film allows gas exchange, which is assumed to prevent ethylene accumulation (which may inhibit shoot expansion) and to reduce humidity in the culture vessels (which prevents anther dehiscence and consequently seed set).

5. Seed set is also sensitive to the presence of kanamycin in the shoot elongation medium, and so antibiotics should be omitted at this stage.

6. Do not put more than two or three shoots in one polypot, as this acts to increase humidity and prevent seed set.

7. Primary transformants (T_1 plants) may set only relatively few seeds *in vitro*, typically about 100 seeds. Germinability of T_1 seeds (T_2 seedlings) is also commonly poor (a typical viability is 70%, but this is variable between transgenic lines). We observe a higher frequency of T_2 seedling survival if seeds are germinated on perlite rather than directly in compost or on agar medium.

8. Aracon® tubes allow individal transgenic lines to be grown up separately in close proximity, without the danger of cross-fertilization or of seed loss.

To Protocol 5.6.2: Screening for T-DNA-tagged mutants

9. The level of kanamycin required to distinguish clearly between sensitive and resistant seedling can vary between lines; this is presumably due to position effects influencing the activity *npt*II genes present in each transformant.

10. Kanamycin-sensitive seedlings bleach, and fail to produce either true leaves or roots on kanamycin selection plates.

11. Emasculation and cross-pollination of *Arabidopsis* flowers is carried out under a dissecting microscope.

12. It is a good idea to repeat the pollination approximately 24 h later just in case the stigma was not fully competent to accept pollen initially.

To Protocol 5.4: Assay of β-glucuronidase (GUS) activity by fluorimetry

13. Do not freeze tissue samples in the GUS buffer as this results in significant loss of enzyme activity.

14. The Na_2CO_3 stops the enzyme reaction and enhances the fluorescence.

To Protocol 5.6: Inverse PCR amplification of T-DNA flanking sequences

15. If it proves to be impossible to segregate out multiple T-DNAs within a particular tagged line, there is no option other than to isolate each T-DNA plus flanking regions individually. The best way to do this is to construct a genomic library from the tagged line and screen using a T-DNA fragment as a probe. Each clone must then be analysed to determine which confers the observed phenotype or GUS staining pattern. In a mutant line the wild-type allele corresponding to each clone must be reintroduced into the tagged line to check for genetic complementation of the mutant phenotype. In a promoter trap line, each 5′ genomic *gus*A fusion fragment is introduced into wild-type plants which are then screened for the original staining pattern.

16. From our work we have some evidence that suggests integration may frequently occur in introns (unpublished data) and, therefore, it is advisable that amplified fragments are used to screen, in the first instance, a genomic library rather than a cDNA library.

17. To amplify genomic sequences flanking the T-DNA left border, we have used the *nested 2* and *nested 3* primers, followed by a confirmatory reamplification

with one of the *LB* primers and *nested* 3. We have found that it is common for deletions (of varying sizes) to occur at the left border on T-DNA integration and, therefore, three left border primers were designed which are sited progressively further from the left border sequence. These primers are also useful for sequencing the T-DNA/genomic flanking region. For right border flanking sequences, we have used *nested 2* and *npt*II primers.

18. The concentration of DNA and Mg^{2+} used in the amplification reaction is critical and it is advisable to test empirically a range of concentrations of each in preliminary experiments.

19. For other primers than those described here, the cycling conditions must be optimized. The extension time is estimated to be 1 min per 1 kb of expected amplification product.

20. If feasible, use a test plasmid at single copy levels mixed with untransformed genomic DNA as a positive control to optimize the reaction conditions.

Acknowledgements

We are grateful to BBSRC and CEC for financial support for our work on gene tagging in *Arabidopsis*.

References

AHMAD, M. and CASHMORE, A.R., 1993, The *HY*4 gene involved in blue light sensing in *Arabidopsis thaliana* encodes a protein with the characteristics of a blue light photoreceptor, *Nature*, **366**, 162–166.

BECHTOLD, N., ELLIS, J. and PELLETIER, G., 1993 *In planta Agrobacterium* mediated gene transfer by infiltration of adult *Arabidopsis thaliana* plants. *Comptes rendus de l'Academie de Sciences de Paris, Sciences de la Vie*, **316**, 1194–1199.

BEVAN, M.W., 1984, Binary *Agrobacterium* vectors for plant transformation, *Nucleic Acids Research*, **22**, 8711–8721.

CALLOS, J.D., DIRADO, M., XU, B., BEHRINGER, F.J., LINK, B.M. and MEDFORD, J.I., 1994, The forever young gene encodes an oxidoreductase required for proper development of the *Arabidopsis* shoot apex, *Plant Journal*, **6**, 835–847.

CASTLE, L.A. and MEINKE, D.W., 1994, A *FUSCA* gene of *Arabidopsis* encodes a novel protein essential for plant development, *Plant Cell*, **6**, 25–41.

CHANG, S.S., PARK, S.K., KIM, B.C., KANG, B.J., KIM, D.U. and NAM, H.G., 1994, Stable genetic transformation of *Arabidopsis thaliana* by *Agrobacterium* inoculation *in planta*, *Plant Journal*, **5**, 551–558.

CHURCH, G.M. and GILBERT, W., 1984, Genomic sequencing, *Proceedings of the National Academy of Sciences (USA)*, **81**, 1991–1995.

CLARKE, M.C., WEI, W. and LINDSEY, K., 1992, High frequency transformation of *Arabidopsis thaliana* by *Agrobacterium tumefaciens*, *Plant Molecular Biology Reporter*, **10**, 178–189.

DAMM, B. and WILLMITZER, L., 1991, *Arabidopsis* protoplast transformation and regeneration. in Lindsey, K. (ed.), *Plant Tissue Culture Manual: Fundamentals and Applications*, Dordrecht: pp. A7: 1–20, Kluwer Academic Publishers.

DEN BOER, B., MONTAGU, M., VAN JOFUKU, K.D. and OKAMURO, J.K., 1991, Cloning of the temperature sensitive floral regulatory gene apetala2-1 from *Arabidopsis thaliana*. Abstract presented at the Keystone meeting on the Genetic Dissection of Plant Processes, Keystone, CO.

DENG, X.-W., MATSUI, M., WEI, N., WAGNER, D., CHU, A.M., FELDMANN, K.A. and QUAIL, P.A., 1992, *COP1*, an *Arabidopsis* regulatory gene, encodes a protein with both a zinc-binding motif and a Gb homologous domain, *Cell*, **71**, 791–801.

FELDMANN, K.A. and MARKS, M.D., 1987, *Agrobacterium*-mediated transformation of germinating seeds of *Arabidopsis thaliana*: a non-tissue culture approach. *Molecular and General Genetics*, **208**, 1–9.

FELDMANN, K.A., MARKS, M.D., CHRISTIANSON, M.L. and QUATRANO, R.S., 1989, A dwarf mutant of *Arabidopsis* generated by T-DNA insertion mutagenesis, *Science*, **243**, 1351–1354.

FOBERT, P.R., MIKI, B. L. and IYER, V. N., 1991, Detection of gene regulatory signals in plants revealed by T-DNA-mediated fusions. *Plant Molecular Biology*, **17**, 837–851.

FORSTHOEFEL, N.R., WU, Y., SCHULZ, B., BENNETT, M.J. and FELDMANN, K.A., 1992, T-DNA insertion mutagenesis in *Arabidopsis*: prospects and perspectives, *Australian Journal of Plant Physiology*, **19**, 353–366.

FROHMAN, M.A., 1990, RACE: Rapid Amplification of cDNA ends, in Innis, M.A., Gelfland, D.H., Shinsky, J.J. and White, T.J. (Eds) *PCR Procotols. A Guide to Methods and Applications*, pp. 28–38, San Diego: Academic Press.

GAMBORG, O.L., MILLER, R.A. and OJIMA, K., 1968, Nutrient requirements of suspension cultures of soybean root cells, *Experimental Cell Research*, **50**, 151–158.

GODDIJN, O.J.M., LINDSEY, K., VAN DER LEE, F.M., KLAP, J.C. and SIJMONS, P.C., 1993, Differential expression in nematode-induced feeding structures of transgenic plants harbouring promoter-*gus*A fusion constructs, *Plant Journal*, **4**, 863–873.

HAYASHI, H., CZAJA, I., LUBENOW, H., SCHELL, J. and WALDEN, R., 1992, Activation of a plant gene by T-DNA tagging: auxin-independent growth *in vitro*, *Science*, **258**, 1350–1353.

HERMAN, P.L. and MARKS, M.D., 1989, Trichome development in *Arabidopsis thaliana*. II. Isolation and complementation of the GLABROUS 1 gene, *Plant Cell*, **1**, 1051–1055.

HEWELT, A., PRINSEN, E., SCHELL, J., VAN ONCKELEN, H. and SCHMÜLLING, T., 1994, Promoter tagging with a promoterless *ipt* gene leads to cytokinin-induced phenotypic variability in transgenic tobacco plants: implications of gene dosage effects, *Plant Journal*, **6**, 879–891.

INNIS, M.A., GELFLAND, D.H., SHINSKY, J.J. and WHITE, T.J., 1990, *PCR Protocols. A Guide to Methods and Applications*, San Diego: Academic Press.

JACOBSEN, S.E. and OLSZEWSKI, N.E., 1993, Mutations at the *SPINDLY* locus of *Arabidopsis* alter gibberellin signal transduction, *Plant Cell*, **5**, 887–896.

JEFFERSON, R.A., KAVANAGH, T.A. and BEVAN, M.W., 1987, GUS fusions: β-glucuronidase as a sensitive and versatile gene fusion marker in higher plants, *EMBO Journal*, **6**, 3901–3907.

KERTBUNDIT, S., DE GREVE, H., DEBOECK, F., VAN MONTAGU, M. and HERNALSTEENS, J.-P., 1991, *In vivo* random β-glucuronidase gene fusions in *Arabidopsis thaliana*, *Proceedings of the National Academy of Sciences (USA)*, **88**, 5212–5216.

KIRCHNER, G., KINSLOW, C.J., BLOOM, G.C. and TAYLOR, D.W., 1993, Nonlethal assay system of β-glucuronidase activity in transgenic roots of tobacco, *Plant Molecular Biology Reporter*, **11**, 320–325.

KONCZ, C., MARTINI, N., MAYERHOFER, R., KONCZ-KALMAN, Z., KORBER, H., RÉDEI, G.P. and SCHELL, J., 1989, High frequency T-DNA-mediated gene tagging in plants, *Proceedings of the National Academy of Sciences (USA)*, **86**, 8467–8471.

LARKIN, J.C., OPPENHEIMER, D.G., POLLOCK, S. and MARKS, M.D., 1993, *Arabidopsis GLABROUS 1* gene requires downstream sequences for function, *Plant Cell*, **5**, 1739–1748.

LEE, I., AUKERMAN, M.J., GORE, S.L., LOHMAN, K.N., MICHAELS, S.D., WEAVER, L.M., JOHN, M.C., FELDMANN, K.A. and AMASINO, R.M., 1994, Isolation of *LUMINIDEPENDENS*: a gene involved in the control of flowering time in *Arabidopsis*, *Plant Cell*, **6**, 75–83.

LINDSEY, K. and JONES, M.G.K, 1989, Stable transformation of sugarbeet protoplasts by electroporation, *Plant Cell Reports*, **8**, 71–74.

LINDSEY, K., WEI, W., CLARKE, M.C., MCARDLE, H.F., ROOKE, L.M. and TOPPING, J.F., 1993, Tagging genomic sequences that direct transgene expression by activation of a promoter trap in plants., *Transgenic Research*, **2**, 33–47.

MANDAL, A., SANDGREN, M., HOLMSTRÖM, K-O., GALLOIS, P. and PALVA, E.T., 1995, Identification of *Arabidopsis thaliana* sequences responsive to low temperature and absisic acid by T-DNA tagging and *in vivo* gene fusion, *Plant Molecular Biology Reporter*, **13**, 243–254.

MARKS, M.D. and FELDMANN, K.A., 1989, Trichome development in *Arabidopsis*. I: T-DNA tagging of the *Glabrous 1* gene, *Plant Cell*, **1**, 1043–1050.

MASCARENHAS, J.P. and HAMILTON, D.A., 1992, Artifacts in the localization of GUS activity in anthers of petunia transformed with a CaMV 35S-GUS construct, *Plant Journal*, **2**, 405–408.

MEYEROWITZ, E.M., 1987, *Arabidopsis thaliana*, *Annual Review of Genetics*, **21**, 93–111.

MURASHIGE, T. and SKOOG, F., 1962, A revised medium for rapid growth and bioassays with tobacco tissue cultures, *Physiologia Plantarum*, **15**, 473–497.

MUSKETT, P., WEI, W. and LINDSEY, K., 1994, Promoter trapping to identify and isolate genes expressed in vascular tissues of *Arabidopsis thaliana*, *Abstracts of the 4th International Conference on Plant Molecular Biology*, 2079, International Society for Plant Molecular Biology.

OCHMAN, H., GERBER, A.S. and HARTL, D.L., 1998, Genetic applications of inverse polymerase chain reaction, *Genetics*, **120**, 621–623.

OKULEY, J., LIGHTNER, J., FELDMANN, K., YADAV, N., LARK, E. and BROWSE, J., 1994, *Arabidopsis FAD2* gene encodes the enzyme that is essential for polyunsaturated lipid synthesis, *Plant Cell*, **6**, 147–158.

ROE, J.L., RIVIN, C.J., SESSIONS, R.A., FELDMANN, K.A. and ZAMBRYSKI, P.C., 1993, The *tusled* gene in *A. thaliana* encodes a protein kinase homolog that is required for leaf and flower development, *Cell*, **75**, 939–950.

ROOKE, L.M., 1995, Studies on position effects and gene tagging in transgenic plants, PhD thesis, University of Leicester, UK.

RUSSELL, J., FULLER, J., WILSON, Z. and MULLIGAN, B., 1991, Protocol for growing *Arabidopsis*, in Flanders, D. and Dean, C. (Eds), *Arabidopsis: The Compleat Guide*, Agricultural and Food Research Council, Plant Molecular Biology *Arabidopsis* Programme.

SAMBROOK, J., FRITSCH, E.F. and MANIATIS, T., 1989, *Molecular Cloning: A Laboratory Manual*, 2nd Edn, Cold Spring Harbor: Cold Spring Harbor Laboratory Press.

SCHMIDT, R. and WILLMITZER, L., 1991, *Arabidopsis* regeneration and transformation (leaf and cotyledon explant system), in: Lindsey, K. (Ed.) *Plant Tissue Culture Manual: Fundamentals and Applications*, pp. A6: 1–17, Dordrecht: Kluwer Academic Publishers.

STOMP, A.-M., 1990, Use of X-Gluc for histochemical localisation of glucuronidase, *Editorial Comments*, United State Biochemical, Cleveland, pp. 5.

TAKAHASHI, T., GASCH, A., NISHIZAWA, N. and CHUA, N.-H., 1995, The *DIMINUTO* gene of *Arabidopsis* is involved in regulating cell elongation. *Genes and Development*, **9**, 97–107.

TOPPING, J.F. and LINDSEY, K., 1995, Insertional mutagenesis and promoter trapping in plants for the isolation of genes and the study of development, *Transgenic Research*, **4**, 291–305.

TOPPING, J.F., WEI, W. and LINDSEY, K., 1991, Functional tagging of regulatory elements in the plant genome, *Development*, **112**, 1009–1019.

TOPPING, J.F., AGYEMAN, F., HENRICOT, B. and LINDSEY, K., 1994, Identification of molecular markers of embryogenesis in *Arabidopsis thaliana* by promoter trapping, *Plant Journal*, **5**, 895–903.

VALVEKENS, D., VAN LIJSEBETTENS, M. and VAN MONTAGU, M., 1992, *Arabidopsis* regeneration and transformation (root explant system), in Lindsey, K. (Ed.) *Plant Tissue Culture Manual: Fundamentals and Applications*, pp. A8: 1–17, Dordrecht: Kluwer Academic Publishers.

VAN LIJSEBETTENS, M., VANDERHAEGEN, R. and VAN MONTAGU, M., 1991, Insertional mutagenesis in *Arabidopsis thaliana*: isolation of a T-DNA-linked mutation that alters leaf morphology, *Theoretical and Applied Genetics*, **81**, 277–284.

WALBOT, V., 1992, Strategies for mutagenesis and gene cloning using transposon tagging and T-DNA insertional mutagenesis, *Annual Review of Plant Physiology and Plant Molecular Biology*, **43**, 49–82.

WALDEN, R., HAYASHI, H. and SCHELL, J., 1991, T-DNA as a gene tag, *Plant Journal*, **1**, 281–288.

WALDEN, R., FRITZE, K., HAYASHI, H., MIKLASHEVICHS, E., HARLING, H. and SCHELL, J., 1994, Activation tagging: a means of isolating genes implicated as playing a role in plant growth and development, *Plant Molecular Biology*, **26**, 1521–1528.

WEI, N. and DENG, X.-W., 1993, Characterisation and molecular cloning of COP9, a genetic locus involved in light-regulated development and gene expression in *Arabidopsis*, in Hangarter, R., Scholl, R., Davis, K. and Feldmann, K., (Eds) *Proceedings of the 5th International Conference on Arabidopsis Research*, p. 14, The Ohio State University, Columbus, Ohio.

WEI, W., MCARDLE, H. and LINDSEY, K., 1994, pΔgusBin19 T-DNA, *Plant Molecular Biology*, **26**, 1021.

WEI, W., TWELL, D. and LINDSEY, K., 1997, Identification of a novel nucleic acid helicase gene in *Arabidopsis* by promoter trapping, *Plant Journal*, in press.

WILKINSON, J.E., TWELL, D. and LINDSEY, K., 1994, Methanol does not specifically inhibit endogenous β-glucuronidase (GUS) activity, *Plant Science*, **97**, 61–67.

YANOFSKY, M.F., MA, H., BOWMAN, J.L., DREWS, G.N., FELDMANN, K.A. and MEYEROWITZ, E.M., 1990, The protein encoded by the *Arabidopsis* homeotic gene *agamous* resembles transcription factors, *Nature*, **346**, 35–38.

Two-Dimensional Polyacrylamide Gel Electrophoresis-Based Analysis for the Identification of Proteins and Corresponding Genes

GUY BAUW and MARC VAN MONTAGU

6.1 Introduction

Two-dimensional gel electrophoresis is the most powerful protein separation technique and is applicable to soluble as well as membrane proteins. By combining isofocusing with sodium dodecylsulphate–polyacrylamide gel electrophoresis (SDS–PAGE), a high resolution and a high separation capacity is obtained. These properties make two-dimensional polyacrylamide gel electrophoresis (2D–PAGE) the method of choice for the analysis of very complex protein mixtures. More than a thousand radioactively labelled proteins can be resolved and detected in a single two-dimensional gel (O'Farrell, 1975; Garrels, 1979).

Cellular extracts of proteins are frequently subjected to 2D-PAGE. The resulting protein pattern is a reflection of the protein composition of the analysed cells. The intensity of each protein spot in the protein gel is correlated with its presence in the analysed tissue and the expression of its corresponding gene. 2D-PAGE of a cellular extract from a tissue which has been subjected to *in vivo* labelling with a radioactive amino acid gives insight into the translation of genes and/or the cellular turnover of the proteins.

As each protein pattern is a reflection of the protein composition of the analysed tissue, changes in gene expression are relatively easily detected with 2D-PAGE. As such, 2D-PAGE is nowadays routinely used for the detection of proteins whose expression is altered by external stimuli (see Damerval *et al.*, 1991). Potential alterations in protein intensities due to gene mutation, introduction or silencing of genes, are detected. Two-dimensional gel electrophoresis is the only technique which allows the visualization of these changes in gene expression at the protein level in a relatively fast, sensitive and reproducible way. Moreover, the expression level of hundreds of proteins are investigated simultaneously.

The many different proteins separated and visualized in a single two-dimensional gel necessitated the development of sophisticated software for an efficient assimilation and storage of the data and a partially automated comparison of multiple two-dimensional gels (Garrels, 1989). An extensive analysis of the protein composition of different tissues of a species and the protein intensities as a function of externally applied growth conditions, are part of a protein database. Today such a protein database has been established for several human cells (Celis *et al.*, 1990), rat cells (Garrels and Franza, 1989) and *Escherichia coli* (Neidhardt *et al.*, 1989).

Vandekerckhove *et al.* (1985) introduced the principle of protein blotting onto chemical-resistant membranes, allowing the direct identification of the NH_2-terminal sequence of gel-separated proteins. The transfer technique was later refined and new chemical-resistant membranes with a higher protein-binding capacity were introduced (Bauw *et al.*, 1987; Matsudaira, 1987; Vandekerckhove *et al.*, 1987). With these improvements, it became possible to determine directly the NH_2-terminal amino acid sequence of proteins separated by 2D-PAGE (Bauw *et al.*, 1987). The major advantage of the blotting technique is that no additional manipulations are needed once the proteins are bound to the membrane to determine the NH_2-terminal sequence.

Later, proteins immobilized onto a membrane were subjected to a chemical or enzymatic cleavage *in situ* on the membrane (Aebersold *et al.*, 1987; Bauw *et al.*, 1988). After the cleavage, the generated polypeptides are eluted, separated by reversed-phase high-pressure liquid chromatography (HPLC) and the amino acid sequence of some of them is determined. This *in situ* digestion of membrane-bound proteins eliminates the problem of the NH_2-terminally blocked proteins and permits the determination of multiple, internal sequences of a single protein.

Owing to the development of protein blotting onto chemical-resistant membranes such as coated glass-fibre and polyvinylidene difluoride (PVDF) membranes, many new proteins purified by 2D-PAGE were identified (Bauw *et al.*, 1987; Celis *et al.*, 1990). In this way, two-dimensional gel electrophoresis evolved from a purely analytical tool to a tool allowing the identification of the visualized protein.

6.2 Two-dimensional electrophoresis of plant proteins

Since its introduction by O'Farrell (1975), the technical aspects for the performance of 2D-PAGE have undergone slight change. The main modification was the introduction of immobilized ampholines by Bjellqvist *et al.* (1982) instead of the commonly used carrier ampholines. Although the basic technique of 2D-PAGE has changed little, numerous modifications and adaptations have been described, the majority of which involved the pretreatment of the protein sample or the use of different detergents to dissolve the proteins. A comprehensive overview of the use of 2D-PAGE for the analysis of plants proteins can be found in Damerval *et al.* (1991).

Two-dimensional gel electrophoresis, as described here, is identical to the method originally described by O'Farrell (1975). The electrophoresis apparatus for isofocusing and SDS-PAGE are based on those described by O'Farrell (1975). The result of the 2D-PAGE is strongly influenced by the preparation of the proteins. Many problems encountered with 2D-PAGE, particularly when analysing plant proteins, are related to the isolation of the protein sample. Plant tissue generally has a lower protein content, and the presence of phenolic compounds, tannins, lipids, and others, interfere with a clean separation of the proteins. Most of these components bind covalently and/or non-covalently to proteins, altering their charge and behaviour. These reactions are not quantitative, dividing the polypeptide amount into several fractions which results in a row of spots or horizontal streaks on the gel. Furthermore, the presence of lipids or phenolics also causes clotting of the proteins.

Many different extraction methods for plant tissue proteins have been described elsewhere (see Damerval *et al.*, 1991). A good pretreatment of proteins for 2D-gel electrophoresis includes four steps: tissue homogenization, protein extraction (separation from other components), concentration, and solubilization of the protein. All steps are important and will influence the two-dimensional separation. Inefficient grinding of the tissue will lead to loss of protein and to an unreliable protein pattern. The presence of phenolics, lignins and tannins in the protein sample may result in protein modification, insolubility or streaking in the gel. Proteins not completely solubilized will not, or only partially, be found in the resulting two-dimensional gel protein pattern, and consequently, this pattern will not reflect the real protein content of the analysed tissue.

6.3 Protein extraction with phenol

The phenol extraction procedure is adapted from Hurkman and Tanaka (1986). The method has been successfully used for 2D-PAGE of xylem tissue of poplar and various tissues of *Arabidopsis* and tobacco. Either freshly harvested or frozen tissue ($-70°C$) is used for the extraction. If the harvested tissue is stored at $-70°C$, do not grind the tissue before storage because grinding liberates the proteases, which will be enzymatically active upon thawing. To reduce the proteolytic activity of proteases, always keep the extract on ice. It is also possible to add a cocktail of protease inhibitors to the extract solution.

After extraction of the protein, the debris is isolated by centrifugation at 4000 *g*. When a higher speed is used, other particles and membrane vesicles will also be (partially) eliminated from the supernatant. This can result in the (partial) removal of protein spots from the protein pattern. For isolation of precipitated protein, always spin with a maximal force of 4000 *g*. Centrifugation with a higher force or drying too extensively will make it difficult or impossible to dissolve the protein pellet.

During the separation of the phenol and water phases, sometimes a thick (up to 1 cm) interphase occurs. Although this interphase contains much protein, *never* add the interphase to the phenol. Taking the interphase together with the phenol will almost certainly result in poor quality two-dimensional gels. During the second phenol extraction most of the protein present in this interphase is extracted.

PROTOCOL 6.1. PROTEIN EXTRACTION

Protein extraction buffer: 0.7 M sucrose, 0.5 M Tris, 30 mM HCl, 0.1 M KCl and 1% 2-mercaptoethanol.

Method

1. Weigh the tissue (0.5–1 g) and place in a chilled mortar. In the mortar, freeze the tissue with liquid N_2 and immediately grind with a pestle to a fine powder.

2. Add extraction buffer (3 ml) to the powder. Homogenize further to a fine suspension and transfer to a Corex tube. Wash the mortar twice with 1 ml of extraction buffer and transfer

107

this solution to the same Corex tube. Vortex the extract solution thoroughly and keep it at 4°C for at least 30 min.

3. Centrifuge the suspension at 4000 g for 10 min in a swinging bucket rotor. Transfer the supernatant to a new Corex tube and keep at 4°C.

4. Resuspend the pellet in 3 to 5 ml of extraction buffer. The pellet can be further homogenized using a glass Dounce homogenizer. Vortex the suspension thoroughly and keep it at 4°C for 30 min.

5. Centrifuge the suspension again at 4000 g for 10 min. Add the supernatant to the first one.

6. To the combined supernatant, add 5 ml of water-saturated phenol. Mix the phenol and the extract thoroughly and keep on ice for 1 h.

7. Separate the phases by centrifugation in a swinging bucket at 8000 g for 10 min. Transfer the upper phenol phase to a new Corex tube. Add 3 ml of fresh water-saturated phenol to the water phase and the interphase, and repeat the extraction.

8. Combine both phenol phases. Precipitate the proteins from the phenol by addition of 20 ml 0.1 M ammonium acetate dissolved in methanol and place overnight at −20°C.

9. Obtain a protein pellet by centrifugation in a swinging bucket (maximum 4000 g) for 10 min. Pour off the supernatant and dissolve the pellet in 1 ml of water.

10. Add nine volumes of cold acetone and place the solution at −20°C for at least 4 h.

11. Isolate the precipitated protein again by a centrifugation at a maximum of 4000 g in a swinging bucket. Pour off the supernatant and briefly dry the pellet in the air until the smell of acetone has disappeared.

12. Dissolve the protein pellet in the appropriate amount of protein sample buffer and freeze at −20°C.

6.4 Isofocusing, the first dimension

Heating of the urea buffer must be avoided as urea decomposes at increased temperature. By the decomposition of urea, proteins may become carbamylated; the charge of the polypeptide is gradually altered. This is reflected in the formation of a row of protein spots and NH_2-terminal blocking of the proteins. Empirically, we experienced that the proteins are better dissolved by freezing than by warming in the sample buffer.

Initially, the electrical field is gradually increased to give the proteins enough time to penetrate the isofocusing gel and to avoid a massive precipitation of the protein. Partial precipitation, especially when a high protein concentration is used, is difficult to avoid, but most of it will redissolve during the isofocusing.

PROTOCOL 6.2. FIRST DIMENSION, ISOFOCUSING

Stock solutions:

1. 30% acrylamide (2 × crystallized) (Serva, Heidelberg, Germany).

2. 2% bisacrylamide (Serva).

3. 10% Nonidet-P40 (NP40) (BDH, Poole, UK).

4. 10% ammonium persulphate (APS).

5. Protein sample buffer: 9.5 M urea, 2% w/v NP40, 2% ampholines (pH 3.5–10) (Pharmacia, Uppsala, Sweden), 100 mM dithiothreitol (DTT).

6. Overlay buffer: 6 M urea, 2% NP40, 1% ampholines (pH 3.5–10), 100 mM DTT.

Composition of isofocusing gels: 5.5 g urea, 1.3 ml 30% acrylamide, 0.5 ml 2% bisacrylamide, 2 ml 10% NP40, 1.5 ml H_2O, and 0.6 ml ampholine mix (ampholines pH 3.5–10: pH 4–7: pH 5–8 in 4:2:2 proportion).

Method

1. Perform the isofocusing in cylindrical rod gels. The gels are made in glass tubes 15 cm long with an inner diameter of 1.4 mm. To clean the glass tubes after usage, follow the procedure: 24 h in 2 M NaOH, rinse with H_2O, 24 h in 1 M acetic acid, rinse with H_2O, rinse with methanol, and dry.

2. Seal the bottom of the glass tubes with Parafilm®. Place a mark 13 cm from the bottom of the tube to indicate to where the tube must be filled with the gel solution.

3. Prepare the isofocusing gel solution in a flask using the composition described above. Dissolve the urea by slightly warming the solution (maximum 30°C).

4. Start the polymerization by the addition of 15 µl 10% APS and 10 µl N, N, N', N'-tetra-methylethylenediamine (TEMED). Then immediately transfer the gel solution into the glass tubes using a syringe with a long, narrow needle. Cautiously bring the needle to the bottom of the glass tube and gently extrude the gel solution from the syringe. *No* air bubbles should be trapped in the gel solution. Overlay the gel solution with 50 µl H_2O.

5. For a complete polymerization, leave the gels for at least 5 h at room temperature.

6. Cautiously remove the Parafilm® from each glass tube and mount the tubes into the vertical isofocusing apparatus.

7. Rinse the bottom of the isofocusing gel with H_2O and remove potential air bubbles. Immerse the gel bottom in the lower electrode buffer which consists of 10 mM H_3PO_4.

8. Remove the H_2O covering the top of the isofocusing gel and rinse the glass tube above the gel twice with H_2O without disturbing the gel surface. Carefully take away all remaining H_2O from the top of the isofocusing gel with a Hamilton syringe.

9. Form three separate layers, consecutively 20 µl protein sample buffer, 20 µl overlay buffer and 20 mM NaOH, on top of each other on the isofocusing gel. Fill the upper electrode chamber with the electrode buffer (20 mM NaOH). The H_3PO_4-containing chamber is connected to the positive pole and the NaOH one to the negative pole of a power supply. Perform the pre-run by applying, consecutively, 200 V for 15 min, 300 V for 30 min, and 400 V for 1 h.

10. After this pre-run, disconnect the isofocusing apparatus from the power supply and remove the NaOH buffer from the upper chamber. With a syringe remove all liquid from the top of the gel tubes and wash the top twice with H_2O. Carefully take off all remaining H_2O.

11. Thaw protein samples at room temperature, warming the samples if necessary, but not above 35°C. Apply up to 150 µl of the protein sample on top of the isofocusing gel. Overlay the protein sample with 20 µl of overlay buffer and fill the remaining part of the glass tube with fresh NaOH-electrode buffer. Fill the upper chamber with fresh NaOH solution which has been degassed and reconnect the isofocusing apparatus to the power supply.

12. Start the isofocusing at 100 V and increase every 15 min by 100 V until 400 V is reached. Run the gels at 400 V for 18 h.

13. After isofocusing, remove the gel tubes from the apparatus. Either perform the second dimension immediately or store the gels in their tube temporarily (maximum 2–3 days) at −20°C.

6.5 Second dimension: SDS-PAGE and protein staining

As acrylamide and bisacrylamide are toxic compounds, care should be taken when using them, especially when handling the solid form. Only the acrylamide solution is filtered through a 0.4 μm filter. All reagents are stored at 4°C. Because the isofocusing gel is made of a very low acrylamide concentration, it stretches easily. Apply the rod gel on the second dimension without elongating it too much. Always wear gloves when handling the two-dimensional polyacrylamide gel to avoid fingerprints on the gel. Coomassie staining is less sensitive than silver staining, but it is the only method which gives a good indication of the quantity of each detected protein. The quantity of a protein spot is best estimated by comparing its intensity with that of a known amount of a reference protein run simultaneously on the gel.

PROTOCOL 6.3. SECOND DIMENSION, SDS-PAGE

Stock solutions:

1. 30% acrylamide (2 × crystallized) (Serva).
2. 2% bisacrylamide (Serva).
3. 1 M Tris-Cl pH 8.6.
4. 1 M Tris-Cl pH 6.8.
5. 20% SDS.
6. 10% APS.
7. Electrophoresis buffer: 60 g Tris, 144 g glycine, 10 g SDS dissolved in 10 litre of water.
8. Second-dimension equilibration buffer: 60 mM Tris, 1% SDS, 20% glycerol, 50 mM DTT adjusted to pH 6.8 with HCl. Add a small amount of bromophenol blue.

For the composition of separation gels see Table 6.1

Method

1. Always clean the glass plates (20 × 20 cm) immediately after use with detergent, rinse extensively with water, and dry before storage. Before assembling, clean the side of the glass plates which will be in contact with the polyacrylamide gel with ethanol.

2. Between the two glass plates, one front and one back plate (Figure 6.1A), place three 1.5-mm-thick plastic spacers in a U form. Place the spacers 0.5 cm from the border of the glass plates. Hold the formed cassette together with clamps. Completely seal the space between the spacers and the outer border of the plates with melted agarose.

3. Make the separation gel solution in a 100-ml vacuum flask by mixing the different components for the separation gels (see Table 6.1). Add SDS, APS, and TEMED after deaeration of the solution. Gently pour the solution between the glass plates to 2 cm from

Table 6.1   Composition of separation gels (Laemmli, 1970)

	Protein (molecular mass range in kDa)				
	> 70	100	70	50	40
Acrylamide concentration:	7.5%	10.0%	12.5%	15.0%	17.5%
Acrylamide	7.5	10.0	12.5	15.0	17.5
Bisacrylamide	3	3.96	5.1	6	7
Tris-Cl (pH 8.7)	11.25	11.25	11.25	11.25	11.25
H_2O	8	4.5	0.9	–	–
20% SDS	0.15	0.15	0.15	0.15	0.15
10% APS	0.1	0.1	0.1	0.1	0.1
TEMED	0.02	0.02	0.02	0.02	0.02

All volumes are expressed in ml. The total volume is sufficient to make one 1.5-mm-thick gel.

the top of the front plate and then cover the solution with a water layer. Pour the separation gel at least 5 h before use.

4. Make the stacking solution by mixing 1.7 ml 30% acrylamide, 0.6 ml 2% bisacrylamide, 1.25 ml Tris-Cl (pH 6.8), and 5.7 ml H_2O. After deaeration of this solution, add 0.05 ml 20% SDS, 0.05 ml 10% APS, and 0.005 ml TEMED and place 1.5 cm of stacking solution on top of the gel. Overlay the stacking gel with H_2O-saturated isobutanol.

5. Before mounting the polyacrylamide cassette in the electrophoresis apparatus, rinse the top of the stacking gel extensively with H_2O to remove all isobutanol. Remove the water from the top of the gel, and mount the cassette in the electrophoresis apparatus.

6. Thaw the glass tubes with the isofocusing gels at room temperature. Connect a water-filled 5-ml syringe with the basic side of the glass tube. Gently extrude the rod gel from the tube onto a plastic sheet by slowly exerting pressure on the syringe.

7. Place the sheet on the border of the front glass plate and gently push the isofocusing gel against the back plate (Figure 6.1B). Cautiously remove the plastic sheet, keeping the isofocusing gel between the two glass plates. Starting from one side, going gradually to the other, gently press the rod gel between the two plates with a round-shaped spatula, until the isofocusing gel makes contact with the stacking gel over its total length (Figure 6.1C). When pressing the isofocusing gel onto the stacking gel, air bubbles between the gels are systematically removed. To equilibrate the isofocusing gels, apply 1 ml of second-dimension equilibration buffer on the gel.

8. After 10 min of equilibration, fill the upper chamber with electrophoresis buffer. Start the electrophoresis at 30 V. When the migration front (bromophenol blue serves as indicator) has entered the separation gel, the voltage can be increased. Run the gels overnight at a constant voltage of 70 V. Run the electrophoresis until the migration front has left the bottom of the gel.

PROTOCOL 6.4.   COOMASSIE STAINING OF THE GEL

Stock solutions:

1. Coomassie staining solution: 2 g/l Coomassie Brilliant Blue R250 (Serva), 45% methanol, 7% acetic acid.

2. Destaining solution: 10% methanol, 10% acetic acid.

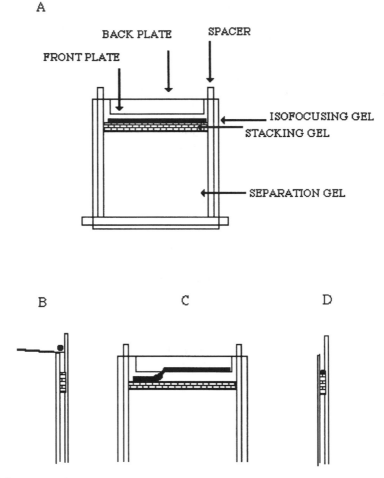

Figure 6.1
A. Assemblage of the second dimension.
B, C, and D, loading of the isofocusing gel on the SDS–PAGE.

Method

1. Remove the polyacrylamide gel from between the glass plates and immerse in 150 ml of Coomassie staining solution. Place on a rotary shaker, and gently shake the gel oversight.

2. Remove the staining solution and rinse the gel extensively with H_2O. Place the gel overnight in the destaining solution on a rotary shaker.

6.6 Protein blotting onto polyvinylidene difluoride membranes

Immobilon® is a very hydrophobic membrane and it is necessary to first use methanol to wet it. The two different sides of the membrane, easily distinguished when wet, have different properties: one side is hydrophilic – water remains equally dispersed; the other is hydrophobic – water tends to form drops. To reduce the occurrence of air bubbles, always place the hydrophilic side of the PVDF membrane facing the polyacrylamide gel. PVDF blots can be stored for years at 4°C after drying in the air but the colour of the spots can fade during storage. At any time, these PVDF blots can be restained with Amido Black after wetting the membrane with methanol.

A complete transfer of proteins from the polyacrylamide gel is only obtained if the gel is blotted immediately. It is possible to transfer Coomassie-stained gels, but the efficiency of transfer is much lower. To blot Coomassie-stained gels, first wash them extensively with distilled water until all acid and methanol is removed. The proteins are remobilized by equilibrating the gel for 2 h in the pre-equilibration buffer. The electrotransfer is then performed as described.

PROTOCOL 6.5.   PROTEIN BLOTTING ONTO IMMOBILON

Stock solutions:

1. Pre-equilibration buffer: 50 mM H_3BO_3, 0.1% SDS adjusted to pH 8 with NaOH.

2. Transfer buffer: 50 mM Tris, 50 mM H_3BO_3.

3. Amido Black staining solution: 1 g/l Amido Black, 40% methanol, 7% acetic acid.

Method

1. After the electrophoresis, place the polyacrylamide gel immediately in 150 ml of pre-equilibration buffer. Allow the gel to swell in this buffer for 1.5 h on a rotary shaker.

2. Meanwhile cut Immobilon® (Millipore, Bradford, MA, USA) to the size of the polyacrylamide gel. First, immerse the PVDF membrane for 5 min in methanol and then wash thoroughly with distilled water to remove all methanol. Place the membrane in water for at least 1 h on a rotary shaker with gentle agitation.

3. To avoid air being trapped between the different layers of the blotting sandwich, assemble the sandwich in a box filled with transfer buffer and pre-wet each part. Make the blotting sandwich by mounting the following components on top of each other: a porous pad, four sheets of Whatman 3MM paper, the equilibrated polyacrylamide gel, the PVDF membrane, four sheets of Whatman paper, and a porous pad. Make sure no air is trapped between two layers.

4. Mount the blotting sandwich into the gel holder unit and insert into the transfer cell (Bio-Rad, Richmond, CA, USA), filled with the transfer buffer. Insert the blotting sandwich into the transfer cell so that the polyacrylamide gel is faced towards the negative electrode and the blotting membrane towards the positive.

5. Carry out the electrotransfer at a constant voltage of 35 V and the transfer will be complete after 8 h. The transfer may also be performed overnight at 30 V.

6. After transfer, disassemble the blotting sandwich and place the PVDF membrane in distilled water and gently agitate for 5 min. Then transfer the membrane into the Amido Black staining solution. After 10 min of gentle agitation in the staining solution, pour off the solution and rinse the membrane with water until the proteins are visible as blue spots on a white background.

6.7 Semi-preparative two-dimensional gel electrophoresis

The maximum amount of protein which can be applied onto a two-dimensional gel depends on many factors: the size of the polyacrylamide gel, the purpose of the separation (analytical versus preparative), the protein detection technique used, and the composition of the protein mixture. The amount of plant material necessary to generate a two-dimensional gel is largely dependent on the tissue. A good quality Coomassie-stained, two-dimensional gel can be generated from 0.1 g leaf or 0.2 g root tissue from *Arabidopsis*, (Figures 6.2 and 6.3), whereas 1 g of xylem tissue of poplar is needed for a comparable two-dimensional gel. On these two-dimensional gels, 100 to 200 µg of protein can be loaded. This estimation is based on extrapolation from the amounts measured by sequence analysis of different protein spots.

The composition of the protein mixture to be analysed influences the total amount that can be loaded on a two-dimensional gel without loss of resolution. Two-dimensional gel electrophoresis of cell suspensions, root or etiolated tissue poses few problems. The total amount of extracted protein is evenly distributed over the different polypeptides. In these gels, it is easy to detect hundreds of different spots, also after Coomassie staining (Figure 6.2). Green tissue is in general more problematic, because most of the protein amount is concentrated in four to five spots (Figure 6.3). To detect many different proteins, more protein must be loaded, resulting in overloading of the prominent spots. This overloading is reflected in the deformed, abnormal spot shape and, in the worst case, horizontal and vertical streaking around the

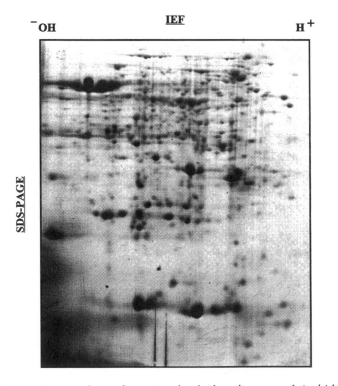

Figure 6.2 Coomassie-stained two-dimensional gel of total extract of *Arabidopsis* root proteins.

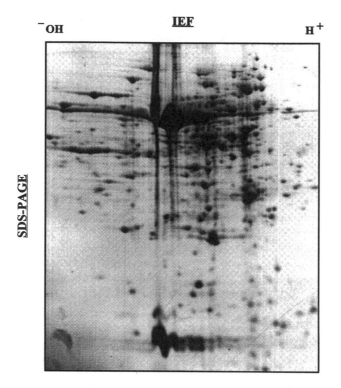

Figure 6.3. Coomassie-stained two-dimensional gel of total extract of *Arabidopsis* siliques. A high protein amount has been applied on the two-dimensional gel to visualize as many proteins as possible. The most prominent protein (ribulose-1,5-bisphosphate carboxylase) is overcharged as the spot is smearing out.

protein spot (Figure 6.3). In this case, the prominent spots can also mask the presence of other proteins and the huge amount of the protein will disturb the local pH gradient.

The highest possible amount of protein is always applied on the two-dimensional gels which will be used for further sequence analysis of the proteins. After electrophoresis, these gels are transferred to PVDF membranes. This blotting step is also a concentration step: the protein is condensed on a surface. The amount of protein applied on semi-preparative gels must not be too large, otherwise the protein spot(s) of interest will overlap with neighbouring proteins after blotting.

6.8 Protein identification of 2D-PAGE-separated proteins by sequence analysis

A first step to sequence analysis is cutting out the protein spots of interest and collecting them from multiple gels until the necessary amount is reached. Only blots on which the protein pattern is completely recognized, the spots of interest are well resolved from the others, and the spots are clearly visible with Amido Black (spot representing more than 0.1 µg), should be used. When the sample will be

115

subjected to an *in situ* digestion with a protease, only the membrane surfaces on which the proteins are bound should be used.

For NH$_2$-terminal sequence analysis, the membrane pieces are immediately applied in the cartridge of the sequencer, which is operated according to the manufacturer's instructions. With the current sequencers, the NH$_2$-terminal sequence analysis is routinely performed on 20 pmol of immobilized protein. In general, 20 amino acids in a row are then determined, if the protein is not NH$_2$-terminally blocked.

Trypsin and Asp-N are the most frequently used proteases for *in situ* digestion, although others like chymotrypsin, pepsin and thermolysin can also be used. Because of the additional manipulations, a higher amount of immobilized protein is required for a successful determination of multiple amino acid sequences. Moreover, the extent of the digestion is highly dependent on the denaturation and accessibility of the immobilized protein. In general, a minimum amount of 50 pmol of protein ought to be used.

Determined amino acid sequences have two applications: identification of the sequenced protein by homology search and basic information for the targeted isolation of the corresponding gene. Comparison of the amino acid sequences with the available protein databanks (PIR, SWISS-PROT) and DNA databanks (GenBank, EMBL) can lead to the identification of the sequenced protein by similarity with already characterized proteins or isolated genes. Nowadays, in many cases, significant similarity will also be found with expressed sequence tags from existing sequencing programmes (Uchimiya *et al.*, 1992; Höfte *et al.*, 1993).

The availability of amino acid sequences of a protein of interest allows a straightforward isolation of corresponding cDNAs or genes. Corresponding degenerate oligonucleotides are synthesized from the most suitable amino acid sequences. Regions with less degeneration (rich in one or two codon-encoded amino acids) are chosen especially, whereas regions containing Leu, Ser and Arg (encoded by six codons) are avoided. These synthetic oligonucleotides are used in two approaches; first, they are used for direct screening of cDNA libraries; and second, two oligonucleotides, one in sense and one in antisense orientation, are used as templates in a polymerase chain reaction (PCR) with first-strand cDNA or genomic DNA. The isolated PCR fragment is then used to screen cDNA or genomic libraries.

PROTOCOL 6.6. *IN SITU* TRYPSIN DIGESTION ON PVDF

Stock solutions:

1. Digestion buffer: 0.1 M Tris-Cl pH 8.5.
2. 0.1 M CaCl$_2$.
3. 5% reduced Triton X-100 (Aldrich, Steinheim, Germany).

Method

1. Cut out the protein spots of interest from the blots with a scalpel and collect in an Eppendorf tube. Only cut out the membrane area to which the protein is bound.
2. If necessary trim the membrane area to pieces of 2 × 2 mm. Add methanol (100 µl) to the Eppendorf tube and after 5 min discard the methanol solution. To remove all remaining methanol, wash the membrane pieces five times with 200 µl of distilled water.

116

3. To the membrane pieces, add 100 μl of digestion buffer and 10 μl of 5% reduced Triton X-100 and leave for 2 h at room temperature.

4. Make a fresh solution of 1 mg/ml porcine trypsin (Sigma, St. Louis, MO, USA) in digestion buffer. To each Eppendorf tube, add 5 μl of 0.1 M $CaCl_2$ and 1 or 2 μl of the trypsin solution. Incubate the digestion at 37°C for at least 6 h.

5. After digestion, transfer the digestion solution to a new Eppendorf tube. Wash the membrane pieces successively with 200 μl 6% trifluoroacetic acid (TFA) and 200 μl H_2O. Add these wash solutions to the digestion solution. Store the peptide solution at −20°C until HPLC analysis.

6. Separate the peptides by reversed-phase HPLC on a C4 column (Vydac Separation Group, Hesperia, CA, USA). Equilibrate the column with solvent A (0.1% TFA) and maintain for 5 min after injection of the peptide solution. Elute peptides in a linear gradient to 55% solvent B (70% acetonitrile, 0.1% TFA) for 50 min. The column eluate is monitored by UV absorption at 214 nm and the peptides are collected manually in an Eppendorf tube.

Acknowledgements

This work was supported by a grant from the Belgian Programme on Interuniversity Poles of Attraction (Prime Minister's Office, Science Policy Programming, #38). G.B. is indebted to the Vlaams Instituut voor de Bevordering van het Wetenschappelijk-Technologisch Onderzoek in de Industrie for a post-doctoral fellowship.

References

AEBERSOLD, R.H., LEAVITT, J., SAAVEDRA, R.A., HOOD, L.E. and KENT, S.B.H., 1987, Internal amino acid sequence analysis of proteins separated by one- and two-dimensional gel electrophoresis after *in situ* protease digestion on nitrocellulose, *Proceedings of the National Academy of Sciences (USA)*, **84**, 6970–6974.

BAUW, G., DE LOOSE, M., INZÉ, D., VAN MONTAGU, M. and VANDEKERCKHOVE, J., 1987, Alterations in the phenotype of plant cells studied by NH_2-terminal amino acid-sequence analysis of proteins electroblotted from two-dimensional gel-separated total extracts, *Proceedings of the National Academy of Sciences (USA)*, **84**, 4806–4810.

BAUW, G., VAN DEN BULCKE, M., VAN DAMME, J., PUYPE, M., VAN MONTAGU, M. and VANDEKERCKHOVE, J., 1988, Protein electroblotting on polybase-coated glass-fiber and polyvinylidene difluoride membranes: an evaluation, *Journal of Protein Chemistry*, **7**, 194–196.

BJELLQVIST, B., EK, K., RIGHETTI, P.G., GIANAZZI, E., GÖRG, A., WESERMEIER, R. and POSTEL, W., 1982, Isoelectric focusing in immobilized pH gradients: principle, methodology and some applications, *Journal of Biochemical and Biophysical Methods*, **6**, 317–339.

CELIS, J.E., CRÜGER, D., KIIL, J., DEJGAARD, K., LAURIDSEN, J.B., RATZ, G.P., BASSE, B., CELIS, A., RASMUSSEN, H.H., BAUW, G. and VANDEKERCKHOVE, J., 1990, A two-dimensional gel protein database of noncultured total human epidermal keratinocytes: identification of proteins that are strongly up-regulated in psoriatic epidermis, *Electrophoresis*, **11**, 242–254.

DAMERVAL, C., ZIVY, M., GRANIER, F. and DE VIENNE, D., 1991, Two-dimensional electrophoresis in plant biology, in Chrambach, A., Dunn, M.J. and Radola, B.J. (Eds) *Advances in Electrophoresis*, Vol. 2, pp. 265–339, Weinheim: VCH Verlagsgesellschaft.

GARRELS, J.I., 1979, Two-dimensional gel electrophoresis and computer analysis of proteins synthesized by clonal cell lines, *Journal of Biological Chemistry*, **254**, 7961–7977.

1989, The QUEST system for quantitative analysis of two-dimensional gels, *Journal of Biological Chemistry*, **264**, 5269–5282.

GARRELS, J.I. and FRANZA, B.R. JR, 1989, Transformation-sensitive and growth-related changes of protein synthesis in REF52 cells. A two-dimensional gel analysis of SV40-, adenovirus-, and Kirsten murine sarcoma virus-transformed rat cells using the REF52 protein databases, *Journal of Biological Chemistry*, **264**, 5299–5312.

HÖFTE, H., DESPREZ, T., AMSELEM, J., CHIAPELLO, H., CABOCHE, M., MOISAN, A., JOURJON, M.-F., CHARPENTEAU, J.-L., BERTHOMIEU, P., GUERRIER, D., GIRAUDAT, J., QUIGLEY, F., THOMAS, F., YU, D.-Y., MACHE, R., RAYNAL, M., COOKE, R., GRELLET, F., DELSENY, M., PARMENTIER, Y., DE MARCILLAC, G., GIGOT, C., FLECK, J., PHILIPPS, G., AXELOS, M., BARDET, C., TREMOUSAYGUE, D. and LESCURE, B., 1993, An inventory of 1152 expressed sequence tags obtained by partial sequencing of cDNAs from *Arabidopsis thaliana*, *Plant Journal*, **4**, 1051–1061.

HURKMAN, W.J. and TANAKA, C.K., 1986, Solubilization of plant membrane proteins for analysis by two-dimensional gel electrophoresis, *Plant Physiology*, **81**, 802–806.

LAEMMLI, U.K., 1970, Cleavage of structural proteins during the assembly of the head of bacteriophage T4, *Nature*, **227**, 680–685.

MATSUDAIRA, P., 1987, Sequence from picomole quantities of proteins electroblotted on polyvinylidene difluoride membranes, *Journal of Biological Chemistry*, **262**, 10035–10038.

NEIDHARDT, F.C., APPLEBY, D.B., SANKAR, P., HUTTON, M.E. and PHILLIPS, T.A., 1989, Genomically linked cellular protein databases derived from two-dimensional polyacrylamide gel electrophoresis, *Electrophoresis*, **10**, 116–122.

O'FARRELL, P.H., 1975, High resolution two-dimensional electrophoresis of proteins, *Journal of Biological Chemistry*, **250**, 4007–4021.

UCHIMIYA, H., KIDOU, S.-I., SHIMAZAKI, T., AOTSUKA, S., TAKAMATSU, S., NISHI, R., HASHIMOTO, H., MATSUBAYASHI, Y., KIDOU, N., UMEDA, M. and KATO, A., 1992, Random sequencing of cDNA libraries reveals a variety of expressed genes in cultured cells of rice (*Oryza sativa* L.), *Plant Journal*, **2**, 1005–1009.

VANDEKERCKHOVE, J., BAUW, G., PUYPE, M., VAN DAMME, J. and VAN MONTAGU, M., 1985, Protein-blotting on Polybrene-coated glass-fiber sheets: a basis for acid hydrolysis and gas-phase sequencing of picomole quantities of protein previously separated on sodium dodecyl sulfate/polyacrylamide gel, *European Journal of Biochemistry*, **152**, 9–19.

VANDEKERCKHOVE, J., BAUW, G., VAN DAMME, J., PUYPE, M. and VAN MONTAGU, M., 1987, Protein-blotting from SDS-polyacrylamide gels on glass-fiber sheets coated with quaternized ammonium polybases, in Walsh, K. A. (Ed.) *Methods in Protein Sequence Analysis*, pp. 261-275, Clifton: Humana Press.

Analysis of Isolated Genes

EKKEHARD HANSEN

7.1 Introduction

Following identification and isolation of differentially expressed genes their characteristics need to be analysed and their expression patterns verified. The method and combination of methods which will be employed will depend very much on the particular interest in the isolated gene. In this chapter a number of basic and more specific methods will be reviewed that will increase the understanding of the gene in question. The scope that various methods offer and the answers they might provide will be discussed, and references are supplied. This should enable the researcher to choose the methodology that best addresses his or her problem.

7.2 Obtaining and analysing the complete gene

7.2.1 Southern analysis

A very important first step in the analysis of the newly isolated gene is the confirmation of the origin of the gene. Especially in experiments that use either non-sterile plant material as a primary source or analyse gene expression induced upon pathogen infection, it is necessary to obtain certainty of the plant origin of the gene in question. This is best done by Southern analysis (Southern, 1975), where the genomic DNA (normally 5–10 µg) of the plant is digested with a series of restriction endonucleases (single or double restrictions) and the fragments are separated according to their size by agarose gel electrophoresis. Following transfer (Southern blotting) to a nylon or nitrocellulose membrane, the genomic plant DNA is hybridized to a labelled probe of the cDNA sequence in question (Southern hybridization). The labelling can be radioactive (^{32}P) or non-radioactive, with a number of very sensitive labelling systems available from a range of suppliers. Southern analysis may also provide information about the copy number of genes and the existence of gene families. Standard protocols can be found in many molecular biology manuals (e.g., Sambrook *et al.*, 1989; Glover and Hames, 1995).

7.2.2 DNA sequencing and sequence analysis

Obtaining the nucleotide sequence of isolated genes is an important step in determining their identity and analysing their character. As the vast majority of genes have homologues in other organisms, comparison with already published sequences, stored in databases, can reveal the function of a particular gene. For *Homo sapiens*, *Arabidopsis thaliana*, *Saccharomyces cerevisiae* and *Caenorhabditis elegans*, internationally co-ordinated programmes to obtain the complete sequences of the respective genomes are under way, generating a wealth of useful information. If no homologues exist in the nucleic acid database, computerized translation of coding sequences into amino acid sequences, and subsequent comparison of certain motifs with those stored in protein databases, can help to elucidate possible functions.

7.2.2.1 *Sequencing*

There are two sequencing methods that are most widely used in the laboratories; the enzymatic, chain termination method developed by Sanger *et al.* (1977), and the chemical degradation method introduced by Maxam and Gilbert (1977). The first method involves the synthesis of a DNA strand by a primer-dependent DNA polymerase *in vitro* using a single strand DNA template. Synthesis is terminated by the inclusion of dideoxynucleoside triphosphates ($2'$,$3'$-ddNTPs) which lack the $3'$-OH group necessary for continued polymerization. An [α-S^{35}]-labelled dNTP is included in the reactions to allow visualization of separated fragments by autoradiography. Four separate reactions each containing either labelled ddATP, ddCTP, ddGTP or ddTTP, are performed for each sample. This allows chain termination in a base-specific manner dependent upon which ddNTP is present. Resulting chain terminated fragments are separated by polyacrylamide gel electrophoresis, and can be read following exposure to X-ray film.

In contrast, the Maxam–Gilbert method does not involve DNA synthesis, but is based on the chemical degradation of the original DNA. A fragment of DNA is radiolabelled at one end, and is subsequently cleaved in five separate chemical reactions that are specific for one base or type of base. The resulting five radiolabelled DNA populations with different lengths (according to the cleavage sites) are separated in polyacrylamide gel electrophoresis, and the molecules of different lengths are visualized by autoradiography. The methodology and protocols of both methods are described in molecular biology manuals, e.g. Sambrook *et al.* (1989).

Nowadays, the Sanger method, because of its ease and rapidity, is used more commonly in laboratories, and automated sequencers using fluorescent labels for terminating the chain reaction are based on it. If no sequencing equipment is available, it is also possible to send the DNA of interest to commercial services in order to have them sequenced. However, the advantage of the Maxam–Gilbert method is that it determines the nucleotide sequence of the original molecule, not its enzymatically produced copy; this is a very important feature where DNA modifications such as methylation are studied, or where the secondary structure and the interaction of DNA with proteins are the subject of the analyse (see Section 7.2.2.3).

7.2.2.2 *Database analysis*

In order to compare sequences with those stored in databases, it is necessary to access these databases. Permission can be obtained by contacting the suppliers.

The most important ones are in Europe: EMBL Databases, EMBL Nucleotide Sequence Data Library in Heidelberg, Germany; SEQNET SERC Database, Daresbury, UK; in the USA: GenBank Genetics Sequence Data Bank, Los Alamos, NM; Protein Identification Resource (PIR) and NBRF Protein Sequence Databases, Washington, DC; and in Japan: DNA Data Bank of Japan, National Institute of Genetics, Misuina. The sequences can be searched by name, plant origin, or sequence data. However, not every sequence in a database is useful, as sometimes cloning vector sequences are included in a database entry, or the information given is not complete, lacking the gene identity or a description how it was obtained. Hence, care must be taken in interpreting information from a database search.

7.2.2.3 *Genomic sequencing*

In plant genomes, up to 50% of all cytosines in C-G or C-N-G sequences may be methylated which has, as part of a gene silencing mechanism, a direct or indirect consequence on gene expression. During recombinant DNA cloning, information about the state of individual nucleotides, the methylation pattern and the chromatin structure, is lost. Genomic sequencing (Church and Gilbert, 1984) permits studies of *in vivo* methylation, based upon the differential reactivity of 5-methylcytosine (5-meC) and cytosine to chemical modification by hydrazine, with 5-meC being much less reactive than cytosine. Digested genomic DNA is chemically modified by hydrazine treatment, and chemically cleaved with piperidine at modified cytosines in the cytosine-specific reaction as described by Maxam and Gilbert (1977). This results in a nested set of DNA fragments within the gene of interest, with the common end being generated by the restriction enzyme cleavage, and the other end being produced by the cytosine-specific cleavage reaction. Following polyacrylamide gel electrophoresis, transfer to a nylon membrane and probing with a single-stranded, gene-specific probe, the 5-meC residues can be detected as missing bands or gaps when compared with unmethylated control DNA of the same sequence. Refinements of this method, using ligation-mediated PCR to produce autoradiograms with high sensitivity and resolution were introduced by Pfeifer *et al.* (1989) and are reviewed in Hornstra and Yang (1993). Another elegant modification, which provides a positive identification of 5-meC, is described in Frommer *et al.* (1992). Cytosine is converted to uracil, but 5-meC remains non-reactive. Using strand-specific primers, the sequence of interest is PCR amplified yielding two strands where all uracil and thymine residues have been amplified as thymine, and only 5-meC has been amplified as cytosine. The PCR products can be sequenced directly or first cloned and then sequenced, using the Sanger method (see Section 7.2.2.1).

7.2.3 *Isolation of 5' and 3' ends of cDNAs*

One important step in determining the putative amino acid sequence and the possible function of the protein encoded by the isolated cDNA is to obtain the full length cDNA. There are a number of protocols designed to recover the ends of cDNA clones. Very frequently used is a method called RACE (**R**apid **a**mplification of **c**DNA **e**nds) developed by Frohman *et al.* (1988). The strategy of this method is the following: using total mRNA from the tissue of interest, reverse transcription is initiated using a gene-specific primer. To isolate the 5' end, a poly(dA) tail is added to the first-strand reaction products using terminal deoxynucleotide transferase (TdT). Finally, the specific product is amplified by PCR using the gene-specific primer and a generic primer complementary to the homopolymer tail of the cDNA. For the 3' end, the mRNA is reverse transcribed using an adapter primer containing a 3' stretch of poly(dT) that anneals to the poly(A) tail of the mRNA transcripts. PCR amplification is performed using oligonucleotides specific to the adapter primer and one oligonucleotide that is gene-specific. RACE experiments can give a smear of products rather than a single band. Modifications of the RACE method that increase the specificity have been described (Jain *et al.*, 1992) which overcome this problem. Further detailed RACE protocols using poly(dG) tailing are described in Geiger *et al.* (1993).

An alternative PCR method for cloning terminal regions has been presented by Towner and Gärtner (1992). Following generation of first- and second-strand cDNA the resulting double-stranded molecules were blunt ended using T4 DNA polymerase and self-ligated at low DNA concentration. For isolation of the 5' terminus, the circularized DNA was inverse-PCR amplified using a 20-mer primer matching a sequence of the 3' end of the gene and an antisense primer to the available 5' region of the gene. The resulting product included the 5' terminus of the gene and could be subcloned or directly sequenced using the antisense primer.

7.2.4 Isolation of genomic sequences and promoters

The isolation of genomic sequences of a gene and comparison with cDNA sequences provide information about the organization of the gene in introns and exons. Non-transcribed sequences like promoters and termination sequences can be determined and analysed. The isolation of the promoter opens possibilities for the analysis of the temporal and spatial expression patterns of the gene in transgenic systems.

7.2.4.1 *Genomic libraries*

The more traditional approach is to construct a genomic library of the plant in question. A genomic library contains (ideally the total) genomic DNA of the plant, fragmented into a manageable size and packed into cloning vectors. The fragments should be large enough to contain entire genes. Overlapping clones make it possible to 'walk' into neighbouring regions. Screening of this library is carried out with a labelled cDNA probe of the isolated, differentially expressed gene. These days the construction of a library is a routine technique. Many manuals deal with the construction and screening of genomic libraries, and commercial kits are available which provide reliable reagents and methodology. It is even possible to order a library from your plant tissue from one of the biotechnology companies, although this is very costly. The construction of genomic libraries and the screening process are very time consuming. However, genomic libraries are very useful when a number of genes are to be isolated, when the ordered array of DNA fragments covers a large region, and when the position of the isolated gene is of particular interest (genome walking). Sometimes, however, a specific gene cannot be found in the library, usually because of problems with cloning of particular parts of the genome, and then alternative techniques have to be employed. PCR technology provides a variety of methods that can be used to isolate unknown sequences flanking a gene whose sequence is partially known.

7.2.4.2 PCR-based methods

Traditional genome analysis requires the generation of DNA libraries in plasmid, phage, cosmid or YAC vectors. This process is time consuming, and sometimes rearrangement in YAC vectors or the inability to clone certain genome regions were encountered. PCR strategies might overcome these shortcomings of the library approach. No single method can be recommended as ideal or superior, as the sequence and the size of the known region vary with each experiment. New PCR strategies or modifications and improvements of established methods are published weekly, and it is useful to refer to specialist journals for updates on the methodology. Briefly, two methods used to generate unknown flanking sequences will be presented.

7.2.4.3 Inverse PCR

The inverse PCR (IPCR) was developed (Ochman *et al.*, 1988; Triglia *et al.*, 1988) to amplify unknown sequences flanking a characterized stretch of DNA. The technique requires digestion and circularization of template DNA using ligation at a high dilution. Using gene specific primers orientated in directions opposite to those normally employed in the PCR, amplification of the the unknown flanking sequences is carried out followed by amplification with gene-specific primers extending into the unknown regions. Problems with this method might arise from difficulties in circularizing the template DNA. Detailed methods can be found in Silver (1991).

7.2.4.4 Vectorette PCR

Vectorette PCR (Arnold and Hodgson, 1991) is a method for generating new sequence information in a known direction starting from a known sequence. The method is a modification of linker-ligated PCR as described by Riley *et al.* (1990). Detailed protocols are described in Allen *et al.* (1994) and Morrison and Markham (1995), the latter for use in mapping the human genome. We have subsequently used this method to isolate the 5' flanking region of an inducible gene in potato (Hansen *et al.*, unpublished data). The duplex vectorette linker consists of two annealed oligonucleotides (50–60-mers) of complementary sequence flanking a non-homologous, constant-mismatched, central region resulting in a 'bubble'. The mismatched portion of the bottom strand contains the same sequence as both the vectorette primer and the sequencing and nested primers. The vectorette also contains a sticky end for ligation. Briefly, genomic DNA is digested with a suitable endonuclease and ligated to the vectorette. The resulting library is used as a template in the PCR. A primer specific to the known gene sequence is used to initiate extension through the unknown genomic sequence to the flanking vectorette oligonucleotide, resulting in a new vectorette strand which is complementary to the bottom strand of the vectorette unit. This new strand differs from the original vectorette top strand as the vectorette molecule contains the mismatched, central region. A primer which has the same sequence as the bottom strand of the mismatched section (and is hence complementary to the newly synthesized strand) is used in the second round of PCR. This primer anneals to the vectorette section of the newly synthesized strand only and initiates extension by DNA polymerase as far as the gene-specific primer used in the first reaction. Subsequent cycles of standard PCR allow amplification of unknown, gene-specific sequences.

7.2.5 *Location of genes on chromosomes*

Improved methods for the isolation of plant chromosomes (Pijnacker and Ferwerda, 1984) have made it possible to employ an *in situ* hybridization technique in order to study the position of genes on chromosomes. In this method, a labelled probe for the gene in question is hybridized to the isolated plant chromosomes. Traditionally, radioactive labelling has been employed in order to detect low-copy genes on plant chromosomes (Huang *et al.*, 1988; Gustafson *et al.*, 1990). In the late 1980s, non-radioactive labelling and detection procedures were introduced (Raap *et al.*, 1989) that offer higher resolution of the hybridization signals and a faster detection time. A standard protocol for chromosome *in situ* hybridization using non-radioactive, digoxigenin-labelled probes has been described by Hinnisdaels *et al.* (1994). Recently, fluorescence *in situ* hybridization (FISH) has been introduced (Griffor *et al.*, 1991; Ricroch *et al.*, 1992) and is now frequently used as an alternative technique. An improved method of FISH has been presented by Fransz *et al.* (1996) that is sensitive enough to detect small targets ($\geq$1.4 kb) of single-copy genes in plants. Multicolour FISH (Leitch *et al.*, 1991; Schmidt and Heslop-Harrison, 1996) may be used for the detection of several genes simultaneously. Especially for the analysis of transgenic plants, the role of the spatial position of the transgene which may be critical for gene expression and gene silencing can be analysed.

7.2.6 *DNA–protein interactions*

Sequence-specific DNA-binding proteins play central roles in many DNA activating or repressing actions, including transcription, replication and recombination. Numerous proteins have been identified using sensitive biochemical methods, including gel mobility shift assays and footprinting.

7.2.6.1 *Gel mobility shift assay*

The electrophoretic mobility in a non-denaturing gel of free DNA is higher than the mobility of DNA–protein complexes. This characteristic is being used to study the nuclear protein-binding capacity of DNA regions of interest (Freid and Crothers, 1981; Garner and Revzin, 1981). Nuclear extracts are prepared and bound to the radioactively labelled DNA of around 30 to 200 bp. Using non-radioactive competitor DNA, the specificity of the binding reaction can be analysed. The DNA–protein complexes are run on a gel that is later dried and exposed to X-ray film. The gel mobility shift assay can be used for the identification of DNA regions with protein-binding capacity. It is a fast and sensitive assay but does not reveal the nucleotide sequences directly involved in the binding process. Standard protocols can be found in Mikami *et al.* (1994) and Parry and Alphey (1995).

7.2.6.2 *Footprinting*

Footprinting has been successfully used to study plant gene regulation. There are various footprinting methods, differing in the enzymes or chemical reagents used to cleave unprotected, protein-free DNA. DNAase I footprinting is the most commonly used method. The DNA fragments of interest are radioactively labelled at one end, mixed with the nuclear extract, incubated with DNAase I and run on a sequencing gel. Comparison of the DNAase I cutting pattern in the presence and absence of proteins reveals information about the protein-binding site. Protocols can be found in Martino-Catt and Kay (1994) and Parry and Alphey (1995). Instead of DNAase I, exonuclease III can be used. The $5'$ end-labelled DNA is incubated with the protein mixture, and digested with the enzyme from the $3'$ end of the DNA strand. After electrophoresis, the length of the undigested labelled DNA reveals the position of one of the boundaries of the protein-binding site. This method, though more sensitive, can be problematic as the 'protection' can also be a consequence of secondary structures of the DNA.

7.3 Analyis of the expression of the isolated gene

7.3.1 *Northern hybridization*

In order to detect expression levels of a particular gene in different organs or tissues of a plant, the method of hybridization of a labelled DNA probe to different RNA populations fixed to a membrane (Northern hybridization) can be employed. After extraction using standard protocols (e.g. Chapter 1 in this volume), the RNA is first electrophoretically size-separated on a denaturing agarose gel, ensuring that the molecules migrate according to their size and are unaffected by secondary structures. The amount of cellular RNA needed to detect a specific mRNA species depends on its prevalence in the mRNA population. High abundant transcripts can be detected with 10–20 μg of cellular RNA. For rare transcripts, poly[A]$^+$-enriched or -purified RNA preparations of about 1–10 μg are used. Northern hybridization is a very powerful and sensitive technique; however, its limitation is that gene expression levels in different cell types in a mixed tissue, or different tissues within an organ, cannot be distinguished. For this purpose *in situ* hybridization should be employed as an alternative or additional method.

7.3.2 In situ *hybridization*

In situ hybridization of nucleic acid probes to mRNA is a very powerful technique to examine the spatial expression patterns of specific genes in tissues within the complex organization of an organ. The method has been employed to analyse the tissue-specific expression of constitutively expressed genes in mature organs, the developmentally regulated expression of certain genes (e.g. Dietrich *et al.*, 1989; Drews *et al.*, 1992), and the inducibility of genes by external factors such as stress or pathogens (e.g., Van der Eycken *et al.*, 1996). A particular strength of this approach is the detection of differential gene expression in single modified cells (e.g. cells modified by the influence of a feeding pathogen) within a homogenous tissue that might not be detected using Northern hybridization.

The plant tissue of interest is fixed and embedded, and subsequently sectioned. The sections are further treated with acid or protease to make the mRNA more accessible, and incubated with bovine serum albumin or acetylated to reduce non-specific binding of the radiolabelled probes. Usually RNA probes or PCR-derived, single-stranded DNA probes (Hannon *et al.*, 1993) are used. Following hybridization the sections are washed to remove unhybridized probe, and subsequently dehydrated. Exposure to autoradiography film reveals the expression patterns of the gene within the tissue section.

There are many publications and some laboratory manuals describing variations of *in situ* hybridization methods in plants, often based on the methods described by Cox and Goldberg (1988) and Angerer and Angerer (1991). It might be necessary to modify the technique for every tissue or gene sequence used, and a review of different approaches is recommended (e.g. Dietrich *et al.*, 1989; Coen *et al.*, 1990; Jackson, 1992; Wilkinson, 1992; Duck, 1994). In addition, non-radioactive labelling methods have been described using digoxigenin- or biotin-labelled nucleic acid probes (Koltunow *et al.*, 1990; Smith *et al.*, 1990).

7.3.3 *Distinguishing expression patterns in multigene families by PCR*

The expression patterns of the various members of a multigene family are difficult or often impossible to distinguish using hybridization techniques. The polymerase chain reaction (PCR) offers approaches to detect specific transcripts, making use of the few distinguishing features between the family members: (i) differences in the restriction sites of gene families have been used as a distinguishing feature, e.g. potato 4-coumarate CoA ligases (Becker-André *et al.*, 1991) and *Arachis* chitinase gene expression (Herget *et al.*, 1990); (ii) RT-PCR have been used to distinguish the expression of gene families using a common end-labelled primer and a gene-specific primer (Simpson *et al.*, 1992a and b). The amplification products can be sized on sequencing gels, and the appearance or disappearance of certain members of the gene family in various tissues or under various conditions can be determined. It is important to remember to design primers such that RT-PCR amplification products can be distinguished from products that are amplified from contaminating genomic DNA. Primers can be designed to span an intron such that products of genomic and cDNA amplification can be distinguished by size difference. If introns are not present, a 3′ primer containing a dT tract and the extreme 3′ terminus of the mRNA will only amplify cDNA in the presence of genomic DNA.

7.3.4 *Analysis in transgenic systems*

Transgenes allow the study of gene function and, in the case of isolated promotors linked to reporter genes, the study of the regulation of the promoter of interest. Commonly the region up to 2 kb upstream of the transcriptional start site, defined as the promoter, is regarded as sufficient to confer the characteristic expression patterns to a gene (Benfey and Chua, 1989). Reporter genes are used in plant transformation for the analysis of promoter activities, for monitoring the efficiency of the selection system used, and for following the inheritance of foreign genes in subsequent plant generations. The ideal reporter protein should show low or no background activity in plants, should not interfere with the plant cell metabolism, and should be detectable using an inexpensive detection system that is sensitive, quantitative and simple to use. In order to monitor changes (up- and down-regulation) in promotor activities, the protein and the mRNA should not be too stable. The bacterial *uid*A (*gus*A) gene encoding the β-glucuronidase enzyme has been the most popular reporter gene in plants for the past 10 years (Jefferson, 1987; Jefferson *et al.*, 1987). Although the enzyme is generally assumed to be stable (Herrera-Estrella *et al.*, 1994), down-expression of β-glucuronidase has been demonstrated within several hours (Herrera-Estrella *et al.*, 1994; Hansen *et al.*, 1996). Other reporter genes, such as chloramphenicol acetyl transferase (*cat*), the first reporter to be introduced into plant cells (Herrera-Estrella *et al.*, 1983), and luciferase (Ow *et al.*, 1986) have frequently been used. The vectors for transformation also need to contain a selectable marker gene controlled by a promoter constitutively active in plants. The most commonly used marker gene is neomycin phosphotransferase (*npt*II), conferring resistance to the antibiotic kanamycin, for which most plants are sensitive; the most frequently used constitutive promoter is the nopalin synthase (*nos*) promoter. Families of vectors with a range of different markers including hygromycin phosphotransferase (*hpt*), dihydrofolate reductase (*dhfr*), phosphinothricin acetyl transferase (*bar*) and bleomycin resistance (*ble*) have been constructed (e.g. Becker *et al.*, 1992).

There is a large number of reports where transgenes have been used to analyse expression patterns and function of a gene of interest. The production of transgenic plants is a common and most valuable tool for analysing expression and activation patterns of promoters. The transgene activity can be monitored during development and the plant can be challenged with various environmental conditions ranging from temperature and light changes to nutritional stress and pathogen attack.

It should be remembered, however, that not only the promoter, but also enhancers and intronic or exonic 3′ sequences play a role in the quantitative expression of plant genes (Dean *et al.*, 1989; Chinn *et al.*, 1996) or in specific activation events (Elliot *et al.*, 1989; Douglas *et al.*, 1991; Dietrich *et al.*, 1992). The role of post-transcriptional regulation in plants has been reviewed by Sullivan and Green (1993). Furthermore, effects of the chromosomal position of the site of integration of the transgene can lead to a variability of transgene expression (Matzke *et al.*, 1989). Methylation events can completely silence a transgene (Matzke and Matzke, 1995; Meyer, 1995) and, following production of transgenic plants, a larger number of individuals should be analysed in order to obtain a clearer picture of the expression patterns of the transgene.

Transformation can be carried out using one of the established systems that are most widely used: (i) transformations mediated by *Agrobacterium tumefaciens* or *A. rhizogenes*; (ii) the use of particle bombardment for transferring DNA into plant cells or tissues; or (iii) direct gene transfer by electroporation or chemical methods. The choice of method will depend on the plant used (not all plants, especially monocotyledons, are susceptible to transformation with *Agrobacterium*) and the equipment available.

7.3.4.1 Agrobacterium-*mediated transformation*

Agrobacterium-mediated transformation is a natural system of gene transfer from the bacterium to dicotolydons where a part of the bacterial Ti plasmid (Ti = tumour inducing), the T-DNA (transfer DNA), is transferred to wounded plant cells and eventually stably integrated into the plant genome. Two features are necessary for this transfer: the short, conserved, *cis*-acting T-DNA border sequences, and the essential *trans*-acting gene products of the virulence (*vir*) region, also located on the Ti plasmid. The plasmid has a size of approximately 200 kb and for this reason is not very amenable to cloning strategies. For directed genetic engineering purposes, a binary vector system has been established (Hoekema *et al.*, 1983), consisting of two autonomously replicating plasmids: a small binary (shuttle) vector that contains the T-DNA borders, and a disarmed helper Ti plasmid from which the oncogenic sequences have been removed but which supplies the *vir* functions. The binary vectors have origins of replication for *Escherichia coli*, in which the cloning is being carried out, and *A. tumefaciens*, into which they are transferred by electroporation or triparental mating in order to be used in the infection and transformation of the plants. The gene of interest and the selectable marker are inserted into the T-DNA flanked by short, conserved right and left border sequences. The transfer of the T-DNA to the plant is activated by the gene products of the *vir* region of the disarmed Ti plasmid. The transferred DNA normally does not undergo major rearrangement processes, and in most cases is integrated as a single copy. A review of *Agrobacterium*-mediated transformation can be found in Hooykaas and Schilperoort (1992), and standard protocols for transformation in Walkerpeach and Velten (1994). By variation of the culture and regeneration techniques, *Agrobacterium*-mediated transformation of the monocot rice has been reported (Chan *et al.*, 1993), and more recently with very high frequency (Hiei *et al.*, 1994).

In general, with susceptible plants, *A. tumefaciens*-mediated transformation is the easier and cheaper and thus most popular choice for producing transgenic plants. It offers the potential to generate transgenic cells at high frequency, usually without significant problems in plant regeneration. For many purposes it might be sufficient to use a model plant for analysis of promoters and genes *in vivo*. *Arabidopsis* is a fast growing weed with a short generation time of several weeks and has been extensively used for genetic analysis. Topping and Lindsey provide a method for transformation of *Arabidopsis* in this volume (Chapter 5).

7.3.4.2 *Particle bombardment*

The first description of using high velocity microprojectiles to deliver exogenous RNA and DNA into epidermal cells of plants by Klein *et al.* (1987) demonstrated the power and versatility of this method. Particle bombardment involves the use of gold or tungsten particles coated with the DNA of interest accelerated by an explosive or electrical discharge, or by pressurized helium, to enter the target plant tissue. Stable transformation of monocots and legumes, both plant groups difficult to handle in culture, has been obtained using highly regenerable tissues. Particle bombardment has been successfully used to transform, amongst many others, the monocots rice (Cao *et al.*, 1992), barley (Wan and Lemaux, 1994), wheat (Vasil *et al.*, 1992), oats (Somers *et al.*, 1992), maize (Gordon-Kamm *et al.*, 1990; Koziel *et al.*, 1993) and sugar cane (Bower and Birch, 1992), and the legumes soybean (McCabe *et al.*, 1988) and bean (Russell *et al.*, 1993). This method is applicable to almost any plant tissue, and is especially potent because it permits the introduction of foreign genes into the germplasm. A number of different instruments, using, for example, a gunpowder device, electrical discharge or compressed helium as accelerating mechanisms, are currently in use. For laboratories not routinely performing genetic transformation techniques, the cost of such an appartus might be a deterrent.

7.3.4.3 *Direct gene transfer by electroporation and chemical methods*

Direct uptake of naked DNA by plant protoplasts in the presence of polyethylene glycol (PEG) or poly-L-ornithine has for many years been an alternative to *Agrobacterium*-mediated delivery of DNA to plants (Davey *et al.*, 1980; Krens *et al.*, 1982; Paszkowski *et al.*, 1984). More recently, electroporation has been successfully used for plant protoplast transformation in order to analyse transient or stable expression of the transgenes (Fromm *et al.*, 1985; Shillito *et al.*, 1985). The applicability of the two methods to generate transformed plants depends on the successful regeneration of protoplast to plants. This has been achieved for the model plant *Arabidopsis* (Damm *et al.*, 1989) and a range of important crops, for example, rice (Shimamoto *et al.*, 1988; Datta *et al.*, 1990), maize (Omirulleh *et al.*, 1993), potato (Masson *et al.*, 1989), tobacco (Paszkowski *et al.*, 1984) and oilseed rape (Guerche *et al.*, 1987). Successful electroporation, however, does not always require the preparation of protoplast. Untreated, intact cells of maize (Songstad *et al.*, 1993) and wheat (Klöti *et al.*, 1993), or intact cells, pretreated by mechanical wounding or exposure to hypertonic solutions, have been used successfully in electroporation experiments (e.g. Morikawa *et al.*, 1986; Lindsey and Jones, 1987; Dekeyser *et al.*, 1990).

Acknowledgement

The author acknowledges financial support including a special fellowship by FENORTE whilst this chapter was compiled.

References

ALLEN, M.J., COLLICK, A. and JEFFREYS, A.J., 1994, Use of vectorette and subvectorette PCR to isolate transgene flanking sequences, *PCR Methods and Applications*, **4**, 71–75.

ANGERER, L.M. and ANGERER, R.C., 1991, Localization of mRNAs by *in situ* hybridisation, *Methods in Cell Biology*, **35**, 37–71.

ARNOLD, C. and HODGSON, I.J., 1991, Vectorette PCR: a novel approach to genome walking. *PCR Methods and Applications*, **1**, 39–42.

BECKER, D., KEMPER, E., SCHELL, J. and MASTERSON, R., 1992, New plant binary vectors with selectable markers located proximal to the left T-DNA border, *Plant Molecular Biology*, **20**, 1195–1197.

BECKER-ANDRÉ, M., SCHULZE-LEFERT, P. and HAHLBROCK, K., 1991, Structural comparison, modes of expression, and putative *cis*-acting elements of the two 4-coumarate: CoA ligase genes in potato, *Journal of Biological Chemistry*, **266**, 8851–8559.

BENFEY, P.N. and CHUA, N.-H., 1989, Regulated genes in transgenic plants. *Science*, **244**, 174–188.

BOWER, R. and BIRCH, R.G., 1992, Transgenic sugarcane plants via microprojectile bombardment, *Plant Journal*, **2**, 409–416.

CAO, J., DUAN, X., McELROY, D. and WU, R., 1992, Regeneration of herbicide resistant transgenic rice plants following microparticle-mediated transformation of suspension culture cells, *Plant Cell Reports*, **11**, 586–591.

CHAN, M.-T., CHANG, H.-H., HO, S.-L., TONG, W.-F. and YU, S.-M., 1993, *Agrobacterium*-mediated production of transgenic rice plants expressing a chimeric α-amylase promoter/β-glucuronidase gene, *Plant Molecular Biology*, **22**, 491–506.

CHINN, A.M., PAYNE, S.R. and COMAI, L., 1996, Variegation and silencing of the *Heat Shock Cognate 80* gene are relieved by a bipartite downstream regulatory element, *The Plant Journal*, **9**, 325–339.

CHURCH, G.M. and GILBERT, W., 1984, Genomic sequencing, *Proceedings of the National Academy of Sciences (USA)*, **81**, 1991–1995.

COEN, E.S., ROMERO, J.M., DOYLE, S., ELLIOT, R., MURPHY, G. and CARPENTER, R., 1990, *floricaula*: a homeotic gene required for flower development in *Antirrhinum majus*, *Cell*, **63**, 1311–1322.

COX, K.H. and GOLDBERG, R.B., 1988, Analysis of plant gene expression, in Shaw, C.H. (Ed.) *Plant molecular biology: A practical approach*, pp. 1–35, Oxford: IRL Press.

DAMM, B., SCHMIDT, R. and WILLMITZER, L., 1989, Efficient transformation of *Arabidopsis thaliana* using direct gene transfer to protoplasts, *Molecular and General Genetics*, **217**, 6–12.

DATTA, S.K., PETERHANS, A., DATTA, K. and POTRYKUS, I., 1990, Genetically engineered fertile indica-rice recovered from protoplasts, *Bio/technology*, **8**, 736–740.

DAVEY, M.R., COCKING, E.C., FREEMAN, J., PEARCE, N. and TUDOR, I., 1980, Transformation of *Petunia* protoplasts by isolated *Agrobacterium* plasmids, *Plant Science Letters*, **18**, 307–313.

DEAN, C., FAVREAU, M., BOND-NUTTER, D., BEDBROOK, J. and DUNSMUIR, P., 1989, Sequences downstream of translation start regulate quantitative expression of two *Petunia rbcS* genes, *Plant Cell*, **1**, 201–208.

DEKEYSER, R.A., CLAES, B., DE RYCKE, R.M.U., HABETS, M.E., VAN MONTAGU, M.C. and CAPLAN, A.B., 1990, Evaluation of selectable markers for rice transformation, *Plant Cell*, **2**, 591–602.

DIETRICH, R.A., MASLYAR, D.J., HEUPEL, R.C. and HARADA, J.J., 1989, Spatial patterns of gene expression in *Brassica napus* seedlings: identification of a cortex-specific gene and localization of mRNAs encoding isocitrate lyase and a peptide homologous to proteinases, *Plant Cell*, **1**, 73–80.

DIETRICH, R.A., RADKE, S.E. and HARADA, J.J., 1992, Downstream DNA sequences are required to activate a gene expressed in the root cortex of embryos and seedlings, *Plant Cell*, **4**, 1371–1382.

DOUGLAS, C.J., HAUFFE, K.D., ITES-MORALES, M., ELLARD, M., PASKOWSKI, U., HAHLBROCK, K. and DANGL, J.L., 1991, Exonic sequences are required for elicitor and light activation of a plant defense gene, but promoter sequences are sufficient for tissue specific expression, *EMBO Journal*, **10**, 1767–1775.

DREWS, D.N., BEALS, T.P., BUI, A.Q. and GOLDBERG, R.B., 1992, Regional and cell-specific gene expression patterns during petal development, *Plant Cell*, **4**, 1383–1404.

DUCK, N.B., 1994, RNA *in situ* hybridization in plants, in Gelvin, S.B. and Schilperoort R.A. (Eds) *Plant Molecular Biology Manual*, G1, pp. 1–13, Dordrecht: Kluwer Academic Publishers.

ELLIOT, R.C., DICKEY, L.F, WHITE, M.J. and THOMPSON, W.F., 1989, *cis*-acting elements for light regulation of pea ferredoxin I gene expression are located within transcribed sequences, *Plant Cell*, **1**, 691–698.

FRANSZ, P.F., STAM, M., MOTIJN, B., TEN HOOPEN, R., WIEGANT, J., KOOTER, J.M., OUD, O. and NANNINGA, N., 1996, Detection of single-copy genes and chromosome rearrangements in *Petunia hybrida* by fluorescence *in situ* hybridization, *Plant Journal*, **9**, 767–774.

FREID, M. and CROTHERS, D.M., 1981, Equilibria and kinetics of *lac* repressor–operator interactions by polyacrylamide gel electrophoresis, *Nucleic Acids Research*, **9**, 6505–6525.

FROHMAN, M.A., DASH, M.K. and MARTIN, G.R., 1988, Rapid production of full length cDNAs from rare transcripts – amplification using a single gene specific oligonucleotide primer, *Proceedings of the National Academy of Sciences (USA)*, **85**, 8998–9002.

FROMM, M.E., TAYLOR, L.P. and WALBOT, V., 1985, Expression of genes transferred into monocot and dicot plant cells by electroporation, *Proceedings of the National Academy of Sciences (USA)*, **82**, 5824–5828.

FROMMER, M., McDONALD, L.E., MILLAR, D.S., COLLIS, C.M., WATT, F., GRIGG, G.W., MOLLOY, P.L. and PAUL, C.L., 1992, A genomic sequencing protocol that yields a positive display of 5-methylcytosine residues in individual DNA strands, *Proceedings of the National Academy of Sciences (USA)*, **89**, 1827–1831.

GARNER, M.M. and REVZIN, A., 1981, A gel electrophoresis method for quantifying the binding of proteins to specific DNA regions: application to components of the *Escherichia coli* lactose operon regulatory system, *Nucleic Acids Research*, **9**, 3047–3060.

GEIGER, M.J., BULL, M., ECKELS, D.D. and GORSKI, J., 1993, Amplification of complementary DNA from mRNA with unknown 5' ends by one-way polymerase chain reaction, *Methods in Enzymology*, **218**, 321–335.

GLOVER, D.M. and HAMES, B.D., 1995, *DNA Cloning: a Practical Approach, Vol. 1: Core techniques.* Oxford: Oxford University Press.

GORDON-KAMM, W.J., SPENCER, T.M., MANGANO, M.L., ADAMS, T.R., DAINES, R.J., START, W.G., O'BRIEN, J.V., CHAMBERS, S.A., ADAMS, W.R., WILLETTS, N.G., RICE, T.B., MACKEY, C.J., KRUEGER, R.W., KAUSCH, A. P. and LEMAUX, P.G., 1990, Transformation of maize cells and regeneration of fertile transgenic plants, *Plant Cell*, **2**, 603–618.

GRIFFOR, M.C., VODKIN, L.O., SINGH, R.J. and HYMOWITZ, T., 1991, Fluorescence *in situ* hybridization to soybean metaphase chromosomes, *Plant Molecular Biology*, **17**, 101–109.

GUERCHE, P., CHARBONNIER, M., JOUANIN, L., TOURNEUR, C., PASZKOWSKI, J. and PELLETIER, G., 1987, Direct gene transfer by electroporation in *Brassica napus*, *Plant Science*, **523**, 111–116.

GUSTAFSON, J.P., BUTLER, E. and McINTYRE, C.L., 1990, Physical mapping of a low-copy DNA sequence in rye (*Secale cereale* L.), *Proceedings of the National Academy of Sciences (USA)*, **87**, 1899–1902.

HANNON, K, JOHNSTONE, E., CRAFT, L.S., LITTLE, S.P., SMITH II, C.K., HEIMAN, M.L. and SANTERRE, R.F., 1993, Synthesis of PCR-derived, single-stranded DNA probes suitable for *in situ* hybridization, *Analytical Biochemistry*, **212**, 421–427.

HANSEN, E., HARPER, G., McPHERSON, M.J. and ATKINSON, H.J., 1996, Differential expression patterns of the wound-inducible transgene *wun1-uidA* in potato roots following infection with either cyst or root knot nematodes, *Physiological and Molecular Plant Pathology*, **48**, 161–170.

HERGET, T., SCHELL, J. and SCHREIER, P.H., 1990, Elicitor-specific induction of one member of the chitinase gene family in *Arachis hypogaea*, *Molecular and General Genetics*, **224**, 469–476.

HERRERA-ESTRELLA, L., DEPICKER, A., VAN MONTAGU, M. and SCHELL, J., 1983, Expression of chimaeric genes transferred into plant cells using a Ti-plasmid-derived vector, *Nature*, **303**, 209–213.

HERRERA-ESTRELLA, L., LEÓN, P., OLSSO, O. and TEERI, T.H., 1994, Reporter genes for plants, in Gelvin, S.B. and Schilperoort, R.A. (Eds) *Plant Molecular Biology Manual*, C2, pp. 1–32, Dordrecht: Kluwer Academic Publishers.

HIEI, Y., OHTA, S., KOMARI, T. and KUMASHIRO, T., 1994, Efficient transformation of rice (*Oryza sativa* L.) mediated by *Agrobacterium* and sequence analysis of the boundaries of the T-DNA, *Plant Journal*, **3**, 271–282.

HINNISDAELS, S., FARBOS, I., DEL-FAVERO, J., VEUSKENS, J., JACOBS, M. and MOURAS, A., 1994, *In situ* hybridization to plant metaphase chromosomes using digoxigenin labeled nucleic acid sequences, in Gelvin, S.B. and Schilperoort, R.A. (Eds) *Plant Molecular Biology Manual*, G2, pp. 1–13, Dordrecht: Kluwer Academic Publishers.

HOEKEMA, A., HIRSCH, P.R., HOOYKAAS, P.J.J. and SCHILPEROORT, R.A., 1983, A binary plant vector strategy based on separation of the *vir* and T-DNA regions of the *Agrobacterium tumefaciens* Ti plasmid, *Nature*, **303**, 179–180.

HOOYKAAS, P.J.J. and SCHILPEROORT, R.A., 1992, *Agrobacterium* and plant genetic engineering, *Plant Molecular Biology*, **19**, 15–38.

HORNSTRA, I.K. and YANG, T.P., 1993, *In vivo* footprinting and genomic sequencing by ligation-mediated PCR, *Analytical Biochemistry*, **213**, 179–193.

HUANG, P.-L., HAHLBROCK, K. and SOMSSICH, I.E., 1988, Detection of a single-copy gene on plant chromosomes by *in situ* hybridization, *Molecular and General Genetics*, **211**, 143–147.

JACKSON, D., 1992, *In situ* hybridization in plants, in Gurr, S.J., McPherson, M.J. and Bowles, D.J. (Eds) *Molecular Plant Pathology: A Practical Approach*, pp. 163–174, Oxford: IRL Press.

JAIN, R., GOMER, R.H. and MURTAGH, JR., J.J., 1992, Increasing specificity from the PCR-RACE technique, *BioTechniques*, **12**, 58–59.

JEFFERSON, R.A., 1987, Assaying chimeric genes in plants: the GUS gene fusion system, *Plant Molecular Biology Reporter*, **5**, 387–405.

JEFFERSON, R.A., KAVANAGH, T.A. and BEVAN, M.W., 1987, GUS fusions: β-glucuronidase as a sensitive and versatile gene fusion marker in higher plants, *EMBO Journal*, **6**, 3901–3907.

KLEIN, T.M., WOLF, E.D., WU, R. and SANFORD, J.C., 1987, High-velocity microprojectiles for delivering nucleic acids into living cells, *Nature*, **327**, 70–73.

KLÖTI, A., IGLESIAS, V.A., WÜNN, J., BURCKHARDT, P.K., DATTA, S.K. and POTRYKUS, I., 1993, Gene transfer by electroporation into intact scutellum cells of wheat embryos, *Plant Cell Reports*, **12**, 671–675.

KOLTUNOW, A.M., TRUETTNER, J., COX, K.H., WALLROTH, M. and GOLDBERG, R.B., 1990, Differential temporal and spatial gene expression patterns occur during anther development, *Plant Cell*, **2**, 1201–1224.

KOZIEL, M.G., BELAND, G.L., BOWMAN, C., CAROZZI, N.B., CRENSHAW, R., CROSSLAND, L., DAWSON, J., DESAI, N., HILL, M., KADWELL, S., LAUNIS, K., LEWIS, K., MADDOX, D., MCPHERSON, K., MEGHJI, M.R., MERLI, E., RHODES, R., WARREN, G.W., WRIGHT, M. and EVOLA, S.V., 1993, Field performance of elite transgenic maize plants expressing an insecticidal protein derived from *Bacillus thuringiensis*, *Bio/Technology*, **11**, 194–200.

KRENS, F.A., MOLENDIJK, L., WULLEMS, G.J. and SCHILPEROORT, R.A., 1982, *In vitro* transformation of plant protoplasts with Ti-plasmid DNA, *Nature*, **296**, 72–74.

LEITCH, I.J., LEITCH, A.R. and HESLOP-HARRISON, J.S., 1991, Physical mapping of plant DNA sequences by simultaneous *in situ* hybridization of two differently labeled fluorescent probes, *Genome*, **34**, 329–333.

LINDSEY, K. and JONES, M.G.K., 1987, Transient gene expression in electroporated protoplasts and intact cells of sugar beet, *Plant Molecular Biology*, **10**, 43–52.

MARTINO-CATT, S.J. and KAY, S.A., 1994, Optimization of DNase footprinting experiments, in Gelvin, S.B. and Schilperoort, R.A. (Eds) *Plant Molecular Biology Manual*, I2, pp. 1–13, Dordrecht: Kluwer Academic Publishers.

MASSON, J, LANCELIN, D., BELLINI, C., LECERF, M., GUERCHE, P. and PELLETIER, G., 1989, Selection of somatic hybrids between diploid clones of potato (*Solanum tuberosum*) transformed by direct gene transfer, *Theoretical and Applied Genetics*, **78**, 153–159.

MATZKE, M. and MATZKE, A., 1995, How and why do plants inactivate homologous (trans) genes? *Plant Physiology*, **107**, 679–685.

MATZKE, M.A., PRIMIG, M., TRNOVSKY, J. and MATZKE, A.J.M., 1989, Reversible methylation and inactivation of marker genes in sequentially transformed tobacco plants, *EMBO Journal*, **8**, 643–649.

MAXAM, A.M. and GILBERT, W., 1977, A new method for sequencing DNA, *Proceedings of the National Academy of Sciences (USA)*, **74**, 560.

McCABE, D.E., SWAIN, W.F., MARTINELL, B.J. and CHRISTOU, P., 1988, Stable transformation of soybean (*Glycine max*) by particle acceleration, *Bio/Technology*, **6**, 923–926.

MEYER, P., 1995, Understanding and controlling transgene expression, *Trends in Biotechnology*, **13**, 332–337.

MIKAMI, K., TAKASE, H. and IWABUCHI, M., 1994, Gel mobility shift assay, in Gelvin, S.B. and Schilperoort, R.A. (Eds) *Plant Molecular Biology Manual*, I1, pp. 1–14, Dordrecht: Kluwer Academic Publishers.

MORIKAWA, H., IIDA, A., MATSUI, C., IKEGAMI, M. and YAMADA, Y., 1986, Gene transfer into intact plant cells by electroinjection through cell walls and membranes, *Gene*, **41**, 121–124.

MORRISON, J.F.J. and MARKHAM, A.F., 1995, PCR-based approaches to human genome mapping, in McPherson, M.J., Hames, B.D. and Taylor, G.R. (Eds) *PCR 2: A practical approach*, pp. 165–196, Oxford: IRL Press.

OCHMAN, H., GERBER, A.S. and HARTL, D.L., 1988, Genetic applications of an inverse polymerase chain reaction, *Genetics* **120**, 621.

OMIRULLEH, S., ABRAHAM, M., GOLOVKIN, M., STEFANOV, I., KARABAEV, M.K., MUSTARDY, L., MOROCZ, S. and DUDITS, D., 1993, Activity of a chimeric promoter with doubled CaMV 35 S enhancer element in protoplast-derived cells and transgenic plants in maize, *Plant Molecular Biology*, **21**, 415–428.

OW, D.W., WOOD, K.V., DeLUCA, M., DE WET, J.R., HELINSKI, D.R. and HOWELL, S.H., 1986, Transient and stable expression of the firefly luciferase gene in plant cells and transgenic plants, *Science*, **234**, 856–859.

PARRY, H.D. and ALPHEY, L., 1995, The utilization of cloned DNAs to study gene organization and expression, in Glover, D.M. and Hames, B.D (Eds) *DNA Cloning: a Practical Approach, Vol. 1: Core techniques,* pp. 143–192, Oxford: Oxford University Press.

PASZKOWSKI, J., SHILLITO, R., SAUL, M.W., MANDAK, V., HOHN, T., HOHN, B. and POTRYKUS, I., 1984, Direct gene transfer to plants, *EMBO Journal*, **3**, 2717–2722.

PFEIFER, G.P., STEIGERWALD, S.D., MUELLER, P.R., WOLD, B. and RIGGS, A., 1989, Genomic sequencing and methylation analysis by ligation mediated PCR, *Science*, **246**, 810–813.

PIJNACKER, L.P. and FERWERDA, M.A., 1984, Giemsa c-banding of potato chromosomes, *Canadian Journal of Genetics and Cytology*, **26**, 415–419.

RAAP, A.K., HOPMAN, A.H.N. and VAN DER PLOEG, M., 1989, Use of hapten modified nucleic acid probes in DNA *in situ* hybridization, in Bullock G. and Petrusz, P. *Techniques in Immunochemistry*, **4**, 167–197.

RICROCH, A., PEFFLEY, E.B. and BAKER, R.J., 1992, Chromosomal location of rDNA in *Allium. In situ* hybridization using biotin- and fluorescein-labeled probes, *Theoretical and Applied Genetics*, **83**, 413–418.

RILEY, J., BUTLER, R., OGILVIE, D., FINNIEAR, R., JENNER, D., POWELL, S., ANAND, R., SMITH, J.C. and MARKHAM, A.F., 1990, A novel method for the isolation of terminal sequence from yeast artificial chromosome (YAC) clones, *Nucleic Acids Research*, **18**, 2887–2890.

RUSSELL, D.R., WALLACE, K., MARTINELL, B.J. and McCABE, D., 1993, Stable transformation of *Phaseolus vulgaris* via electric discharge mediated particle acceleration, *Plant Cell Reports*, **12**, 165–170.

SAMBROOK, J., FRITSCH, E.F. and MANIATIS, T., 1989, *Molecular Cloning: a Laboratory Manual*, 2nd ed, Cold Spring Harbor: Cold Spring Harbor Laboratory Press.

SANGER, F., NICKLEN, S. and COULSON, A.R., 1977, DNA sequencing with chain terminating inhibitors, *Proceedings of the National Academy of Sciences (USA)*, **74**, 5463–5467.

SCHMIDT, T. and HESLOP-HARRISON, J.S., 1996, High-resolution mapping of repetitive DNA by *in situ* hybridization: molecular and chromosomal features of prominent dispersed and discretely localized DNA families from the wild beet species *Beta procumbens*, *Plant Molecular Biology*, **30**, 1099–1114.

SHILLITO, R.D., SAUL, M.W., PASZKOWSKI, J., MÜLLER, M. and POTRYKUS, I., 1985, High efficiency direct gene transfer to plants, *Bio/Technology*, **3**, 1099–1103.

SHIMAMOTO, K., TERADA, R., IZAWA, T. and FUJIMOTO, H., 1988, Fertile transgenic rice plants regenerated from transformed protoplasts, *Nature*, **338**, 274–276.

SILVER, J., 1991, Inverse polymerase chain reaction, in McPherson, M.J., Quirke P. and Taylor G.R. (Eds) *PCR: A Practical Approach*, pp. 137–146, Oxford: Oxford University Press.

SIMPSON, C.G., SAWBRIDGE, T.I., JENKINS, G.I. and BROWN, J.W.S., 1992a, Expression analysis of multigene families by RT-PCR, *Nucleic Acids Research*, **20**, 5861–5862.

SIMPSON, C.G., SINIBALDI, R. and BROWN, J.W.S., 1992b, Rapid analysis of plant gene expression by a novel reverse transcriptase-PCR method, *Plant Journal*, **2**, 835–836.

SMITH, A.G., GASSER, C.S., BUDELIER, K.A. and FRALEY, R.T., 1990, Identification and characterization of stamen- and tapetum-specific genes from tomato, *Molecular and General Genetics*, **222**, 9–16.

SOMERS, D.A., RINES, H.W., GU, W., KAEPPLER, H.F. and BUSHNELL, W.R., 1992, Fertile, transgenic oat plants, *Bio/Technology*, **10**, 1589–1594.

SONGSTAD, D.D., HALAKA, F.G., DEBOER, D.L., ARMSTRONG, C.L., HINCHEE, M.A.W., FORD-SANTINO, C.G., BROWN, S.M., FROMM, M.E. and HORSCH, R.B., 1993, Transient expression of GUS and anthocyanin constructs in intact maize immature embryos following electroporation, *Plant Cell, Tissue and Organ Culture*, **33**, 195–201.

SOUTHERN, E.M., 1975, Detection of specific sequences among DNA fragments separated by gel electrophoresis, *Journal of Molecular Biology*, **98**, 503–517.

SULLIVAN, M.L. and GREEN, P.J., 1993, Post-transcriptional regulation of nuclear-encoded genes in higher plants: the roles of mRNA stability and translation, *Plant Molecular Biology*, **23**, 1091–1104.

TOWNER, P. and GÄRTNER, W., 1992, cDNA cloning of 5′ terminal regions, *Nucleic Acids Research*, **20**, 4669–4670.

TRIGLIA, T., PETERSON, M.G. and KREMP, D.J., 1988, A procedure for *in vitro* amplification of DNA segments that lie outside the boundaries of known sequence, *Nucleic Acids Research* **16**, 8186.

VAN DER EYCKEN, W., DE ALMEIDA ENGLER, J., INZÉ, D., VAN MONTAGU, M. and GHEYSEN, G., 1996, A molecular study of root-knot nematode-induced feeding sites, *Plant Journal*, **9**, 45–54.

VASIL, V., CASTILLO, A.M., FROMM, M.E. and VASIL, I.K., 1992, Herbicide resistant fertile transgenic wheat plants obtained by microprojectile bombardment of regenerable embryogenic callus, *Bio/Technology*, **10**, 667–674.

WALKERPEACH, C.R. and VELTEN, J., 1994, *Agrobacterium*-mediated gene transfer to plant cells: cointegrate and binary vector systems, in Gelvin, S.B. and Schilperoort, R.A. (Eds) *Plant Molecular Biology Manual*, B1, pp. 1–19, Dordrecht: Kluwer Academic Publishers.

WAN, Y. and LEMAUX, P.G., 1994, Generation of large numbers of independently transformed fertile barley plants, *Plant Physiology*, **104**, 37–48.

WILKINSON, D.G. (Ed.), 1992, *In situ Hybridization: A Practical Approach*, Oxford: IRL Press.

Index